PERGAMON INTERNATIONAL LIBRARY
of Science, Technology, Engineering and Social Studies
The 1000-volume original paperback library in aid of education, industrial training and the enjoyment of leisure
Publisher: Robert Maxwell, M.C.

EXPERIMENTS IN ENVIRONMENTAL CHEMISTRY
A Laboratory Manual

THE PERGAMON TEXTBOOK INSPECTION COPY SERVICE

An inspection copy of any book published in the Pergamon International Library will gladly be sent to academic staff without obligation for their consideration for course adoption or recommendation. Copies may be retained for a period of 60 days from receipt and returned if not suitable. When a particular title is adopted or recommended for adoption for class use and the recommendation results in a sale of 12 or more copies, the inspection copy may be retained with our compliments. The Publishers will be pleased to receive suggestions for revised editions and new titles to be published in this important International Library.

PERGAMON SERIES ON
ENVIRONMENTAL SCIENCE
Series Editors: O. HUTZINGER and S. SAFE

VOLUME 4

Other Titles in the Series

Volume 1
HUTZINGER, VAN LELYVAND & ZOETEMAN
Aquatic Pollutants: Transformation and Biological Effects

Volume 2
ZOETEMAN
Sensory Assessment of Water Quality

Volume 3
ALBAIGES
Analytical Techniques in Environmental Chemistry

Soon to be published

FERGUSSON
Inorganic Chemistry: The Environment and Society

Related Pergamon Journals*

CHEMOSPHERE (Chemistry, Physics, Biology and Toxicology as Focused on Environmental Problems)
ENVIRONMENT INTERNATIONAL (A Journal of Science, Technology, Health, Monitoring and Policy)
ION-SELECTIVE ELECTRODE REVIEWS (Applications, Theory and Development)
PROGRESS IN ANALYTICAL ATOMIC SPECTROSCOPY

*Free specimen copy of any Pergamon journal available on request from your nearest Pergamon office.

EXPERIMENTS IN ENVIRONMENTAL CHEMISTRY

A Laboratory Manual

by

P. D. VOWLES and D. W. CONNELL
School of Australian Environmental Studies,
Griffith University, Australia

PERGAMON PRESS
OXFORD · NEW YORK · TORONTO · SYDNEY · PARIS · FRANKFURT

U.K.	Pergamon Press Ltd., Headington Hill Hall, Oxford OX3 0BW, England
U.S.A.	Pergamon Press Inc., Maxwell House, Fairview Park, Elmsford, New York 10523, U.S.A.
CANADA	Pergamon of Canada, Suite 104, 150 Consumers Road, Willowdale, Ontario M2J 1P9, Canada
AUSTRALIA	Pergamon Press (Aust.) Pty. Ltd., P.O. Box 544, Potts Point, N.S.W. 2011, Australia
FRANCE	Pergamon Press SARL, 24 rue des Ecoles, 75240 Paris, Cedex 05, France
FEDERAL REPUBLIC OF GERMANY	Pergamon Press GmbH, 6242 Kronberg-Taunus, Hammerweg 6, Federal Republic of Germany

Copyright © 1980 P D Vowles and D W Connell

All Rights Reserved. No part of this publication may be reproduced, stored in a retrieval system or transmitted in any form or by any means: electronic, electrostatic, magnetic tape, mechanical, photocopying, recording or otherwise, without permission in writing from the publishers.

First edition 1980

British Library Cataloguing in Publication Data
Vowles, P D
Experiments in environmental chemistry.
- (Pergamon series on environmental science;
vol. 4).
1. Environment chemistry - Experiments
I. Title   II. Connell, D W
574.5   QD43   80-40270
ISBN 0-08-024010-0 Hardcover
ISBN 0-08-024009-7 Flexicover

In order to make this volume available as economically and as rapidly as possible the authors' typescripts have been reproduced in their original forms. This method has its typographical limitations but it is hoped that they in no way distract the reader.

Printed in Great Britain by A. Wheaton & Co. Ltd., Exeter

CONTENTS

PREFACE vii

INTRODUCTION 1

PART 1 BIOCHEMICAL PROCESSES IN AQUATIC SYSTEMS 3

1.1 Photosynthesis, Respiration and Biochemical Oxygen Demand. . . . 4
1.2 Eutrophication 14
1.3 Sewage Treatment - A Field Trip 22

PART 2 TOXIC SUBSTANCES IN THE ENVIRONMENT 29

2.1 Insecticides in Cigarette Smoke 30
2.2 Gas Chromatography of Volatile Hydrocarbons 36
2.3 Toxicity of Copper Ions Toward Aquatic Biota 43
2.4 Lead in Household Paint 51
2.5 Atmospheric Pollutants 54

PART 3 FOOD ADDITIVES AND CONTAMINANTS 63

3.1 Aflatoxins in Peanuts 64
3.2 Food Additives 69
3.3 Erucic Acid Content of Bread 74
3.4 DDT in Human Milk 77

PART 4 CHEMICAL ECOLOGY 83

4.1 Chemical Defence of the Monarch Butterfly 84

PART 5 FIELD SURVEY 91

5.1 Stream Pollution 92

Subject Index 101

PREFACE

At the present there is no single well defined body of knowledge which can be classified as "Environmental Chemistry". However, work has been progressing in this topic since the early origins of science but has been included under such titles as "Ecology", "Biochemistry", "Limnology", etc. In broad terms Environmental Chemistry may be considered to encompass the factors affecting the distribution and interaction of elements and compounds in the environment, their modes of transport and transfer and their effects on biological and other systems.

The emphasis in most courses on Environmental Chemistry has been on the behaviour of substances which have become important in environmental management. Thus attention has focussed on air and water pollution, behaviour of toxic materials, food contaminants and so on. The set of experiments described follows this general pattern. All of the experiments have been used in conjunction with theoretical courses in Environmental Chemistry which the authors have developed at Griffith University.

During the last ten years the growth of interest in Environmental Chemistry has been reflected in the initiation of new courses in Universities and the publication of theoretical textbooks. We believe that a book of experiments is timely since it is in the laboratory and field work component that the teaching of Environmental Chemistry seems to be least developed.

The experiments have been designed to provide the laboratory and field component of a first course in Environmental Chemistry. The subjects which can be included in such an exercise are many but the constraints of time, level of expertise of students, general availability of equipment and facilities place limits on this. The programme has been assembled assuming students have a basic background in chemistry and chemical laboratory procedures.

We extend our thanks to our colleagues, particularly Ib Knudsen, Peter Lather, Greg Miller and George Proud and also our students who have helped us develop this material. We are grateful to Mrs Beverley Mickelburgh for typing the manuscript, and to Gary Worthington for preparing the diagrams.

Griffith University   P.D. Vowles

Brisbane   D.W. Connell

Australia

INTRODUCTION

The activities in this set of experiments provide practical, first hand experience in the observation of chemical processes which occur in the environment. A variety of techniques are used which have wide application in governmental laboratories, industry and research. This experience will provide a sound basis for complementary theoretical courses in Environmental Chemistry.

The objectives of this course of experiments are:

1. To develop a practical understanding of some of the basic chemical processes which occur in the environment.

2. To gain experience in the collection, collation and interpretation of data on environmental pollution.

3. To gain experience in the techniques used in Environmental Chemistry.

4. To develop skills in reporting chemical information on the environment.

The experiments have been divided into five parts. The first part, *Biochemical Processes in Aquatic Systems*, is basically concerned with the transformations of carbon, nitrogen, phosphorus and energy in aquatic systems. This gives a practical basis for discussion of a wide variety of aspects of Environmental Chemistry including photosynthesis, respiration, biogeochemical cycling, primary production, plant nutrients, water quality, eutrophication and waste water-treatment. This part also provides the fundamental knowledge to develop further studies in the biological aspects of aquatic ecosystems.

In Part 2, *Toxic Substances in the Environment*, a wide assortment of environmental contaminants are considered in terms of their behaviour and occurence in various sectors of the environment. This section introduces gas chromatography, atomic absorption spectroscopy, thin layer chromatography, column chromatography and techniques for the measurement of atmospheric contaminants. It also includes a bioassay experiment in which the techniques for measurement of the toxicity of a substance are outlined. On this basis the environmental properties of insecticides, heavy metals, petroleum compounds and atmospheric contaminants can be discussed and developed.

In many ways Part 3 *Food Additives and Contaminants* is similar to Part 2 but is specifically concerned with food and the occurrence of foreign substances which result from deliberate additions or other processes.

The role of chemical substances in controlling and regulating the behaviour of animals and plants in natural ecosystems is a rapidly developing area of research. The relationships uncovered are generally too complex to lend themselves to short term student experiments. Part 4 *Chemical Ecology* thus includes only one experiment which is illustrative of the type of intractions involved in Chemical Ecology. This experiment leads to discussion of allelochemicals, pheromones, chemical defence substances and so on.

Part 5 *Stream Pollution* is an integrative exercise in which a wide variety of information is collated and co-ordinated in relation to stream pollution.

Students will obtain maximum advantage from this work by reading the notes and having a clear understanding of the activity prior to the class. In some cases laboratory and field work involves the use of chemicals and equipment which may be dangerous if mishandled. At all times students should exercise careful judgement and also follow closely laboratory safety procedures.

Each student should keep a laboratory notebook to record raw data such as weights and meter readings, rough calculations, and sketches, which are collected during the activities.

Although there is no fixed format, the following headings should be included in reports on activities:

>Title and Date
>Aim
>Results
>Answers to Questions

There is no need to duplicate the list of materials and methods given in the instructions. If you wish, you may describe significant departures from the given methods. Results should be clearly headed, written, tabulated and graphed as appropriate. The questions connected with the activity should be answered so that all relevant points are covered, but this should be done clearly and concisely.

PART 1

Biochemical Process in Aquatic Systems

1.1 PHOTOSYNTHESIS, RESPIRATION AND BIOCHEMICAL OXYGEN DEMAND

INTRODUCTION

The chemical beginnings of life are thought to have taken place in simple photocatalyzed reactions within the primitive atmosphere of the Earth. By processes of continual combination and decay, the basic types of molecules produced built up over long periods of geological time to produce the Earth's living ecosystem. Today, the photochemical production of living organisms and their death and decay continues in the highly complex cycle of photosynthesis and respiration.

Photosynthesis includes the photochemical and associated processes that occur in green plants and some bacteria following the absorption of light. These processes can be simply represented as the use of light energy with carbon dioxide and water to form glucose and oxygen:

$$6CO_2 + 6H_2O + \text{light energy} \xrightarrow{\text{chlorophyll}} C_6H_{12}O_6 + 6O_2 \quad \text{(Equation 1.1.1)}$$

This reaction involves the chemical *reduction* (addition of hydrogen) of carbon dioxide to produce the carbohydrate and oxygen, with the chlorophyll acting as a catalyst. The organisms which can manufacture organic substances from inorganic compounds and energy in this way are described as "autotrophs".

While Equation 1.1.1 is a reasonably accurate representation of the overall photosynthetic process for the production of glucose, plants also contain a variety of additional substances. Other carbohydrates, such as cellulose, protein and fats, are produced in the photosynthetic manufacture of plants. More complex equations, some involving nitrogen, phosphorus and other substances, are thus required to express the overall reaction in these cases. In all cases the transformation from the reactants, water, carbon dioxide and other substances, to the products, plant matter, is a highly complex process not completely understood.

Perhaps the most obvious and arresting feature of plants are their colours, ranging from green in most terrestrial species to orange and red in aquatic species. Chlorophyll, specifically chlorophyll-*a*, is the dominant pigment and plays the major role in photosynthesis. The absorption of the blue and red components of white light and the reflection of the yellow and green wavelengths gives chlorophyll its characteristic green colour (Fig. 1.1.1).

The presence of other pigments, generally referred to as accessory pigments, in plants is important in that the intermediate wavelengths may be absorbed. These accessory pigments do not participate directly in photosynthesis, but instead collect light energy in the wavelengths not absorbed by chlorophyll-*a* (i.e. yellow and green) and transmit the light energy to chlorophyll-*a*. These accessory pigments are particularly important to aquatic plants since water strongly absorbs the longer wavelength (red) radiation.

The photosynthetic manufacture of plant materials, which can be later used by higher organisms as food, is described as primary production. The amount of primary production is affected by a number of factors such as quantity of solar radiation, efficiency of absorption of solar radiation, availability of water etc. (see Table 1.1.1). Solar energy not absorbed is converted into heat and ultimately irradiated into space.

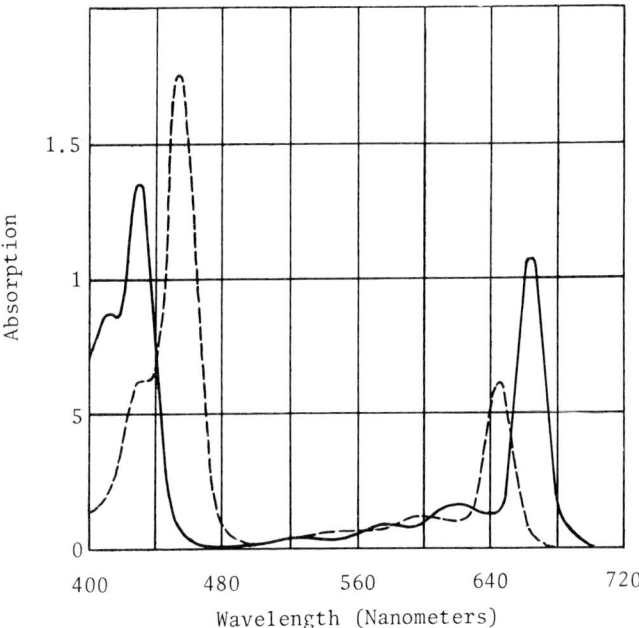

Fig. 1.1.1 These Absorption Spectra show that chlorophyll-a (solid line) and chlorophyll-b (broken line) strongly absorb blue and red light. The green, yellow and orange wavelengths lying between the peaks are reflected and give both pigments their familiar green colour.

It is noteworthy that a limited number of autotrophs may obtain energy from sources other than sunlight energy. In most cases such organisms derive energy from the oxidation of inorganic substances. However, the overall role of these organisms in major ecosystems is insignificant.

Whereas photosynthesis is unique to plants, all organisms respire. Respiration comprises the *oxidation* of carbohydrates and other organic compounds to release the stored energy, usually with the formation of carbon dioxide and water. Glucose is the main foodstuff in respiration, and its overall stoichiometry of reaction is essentially the reverse of Equation 1.1.1.

$$C_6H_{12}O_6 + 6O_2 \rightarrow 6CO_2 + 6H_2O + 673 \text{ kcal} \qquad \text{(Equation 1.1.2)}$$

Apart from carbohydrates, the major oxidizable materials in the natural environment are lipids and proteins. In mammals and a number of other organisms oxygen from air is introduced into a circulatory fluid (blood) through the lungs and thus transmitted to the muscles where major respiration occurs. Many aquatic organisms respire by a somewhat similar process but oxygen is obtained from dissolved oxygen in the water mass by the gills. Microorganisms do not require any special mechanism to obtain oxygen since this can diffuse through the cell wall to the site of consumption.

TABLE 1.1.1 Relationships Between Primary Production and Solar Radiation

Plant	Solar Radiation ($kcal/m^2/day$)	Primary * Production ($kcal/m^2/day$)
sugar cane (Hawaii)	4000	306
irrigated maize (Israel)	6000	405
sugar beet (England)	2650	202

* Gross Primary Production which includes energy used in respiration by the plant.

(From: Odum E.P. (1971), *Fundamentals of Ecology*, 3rd Ed., Saunders, Philadelphia, p. 45).

Fig. 1.1.2 shows the energy transfers which can occur in an aquatic ecosystem involving a wide variety of organisms. The only significant source of energy is in the primary production of plants. This production is consumed in the respiration process by the plants themselves and by organisms in the dependent food web. The oxygen required in this process is derived from the oxygen dissolved in the water mass. Oxygen has limited solubility in both fresh and seawater and under most conditions which are encountered is not present in concentrations greater than approximately 14 mg/l.

Present day land and water uses are such that unusually large quantities of organic matter can be discharged into waterways. This is usually liquid materials such as sewerage discharges and urban run-off water which can be rich in carbohydrate, protein and fat. Bacteria are well adapted to utilise these wastes rapidly and their numbers build up utilising energy derived from respiration (see Equation 1.1.2). The consequent consumption of oxygen can lead to severe depletion of the dissolved oxygen present, often to zero. This has clear implications for the survival of other organisms which also require dissolved oxygen for respiration since a minimum concentration for the survival of fish is generally considered to be 5mg/l.

BIOCHEMICAL PROCESSES IN AQUATIC SYSTEMS

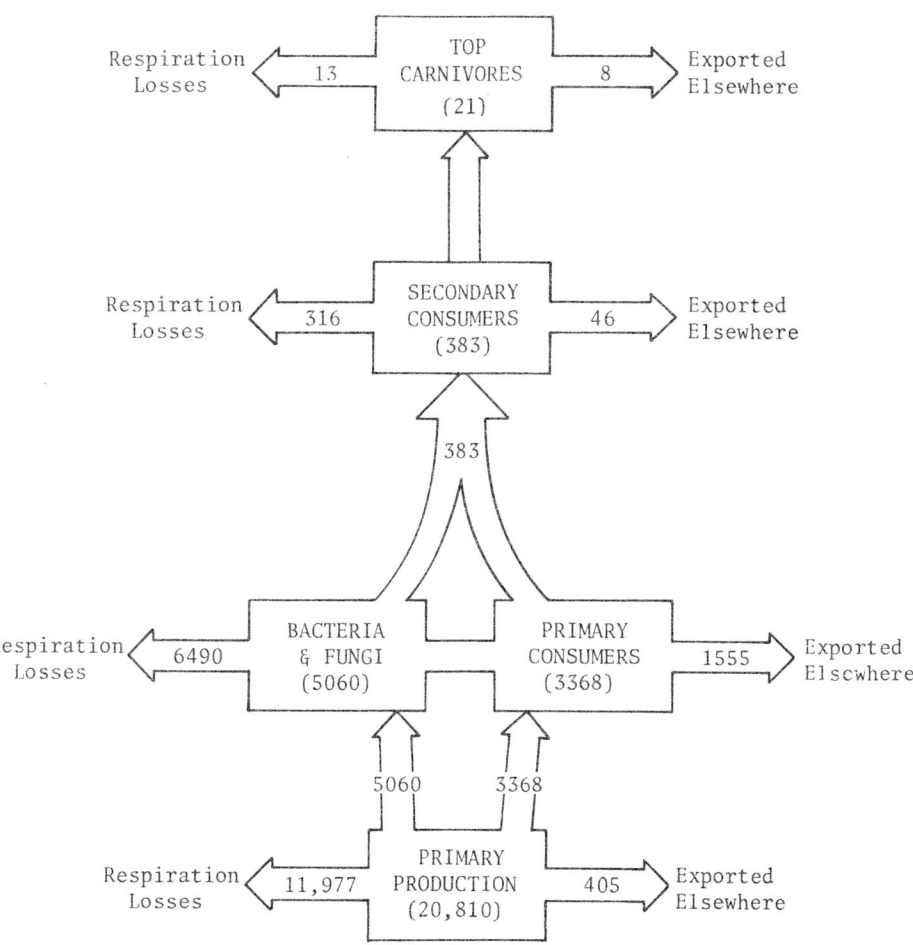

Fig. 1.1.2 Energy Flow and Budget in Silver Springs, Florida (kcal/m^2/year).

(From: Adapted from Philipson, J. *Ecological Energetics*, Institute of Biology, Study in Biology No 1. Edward Arnold, London, 1976, p.15.

It is important in water quality control then, to have a measure of the potential oxygen uptake that accompanies the bacterial consumption of organic material in a waste discharge. This is the basis of the Biochemical Oxygen Demand (BOD) test. In this test, waste waters are diluted, seeded with some common bacteria if necessary, and incubated for a standard time of 5 days to allow aerobic biodegradation to take place. The measured drop in dissolved oxygen over that period is termed the BOD for the particular sample. The BOD of stream water itself can be measured to assess levels of pollution.

In practice, BOD values may range up to 5000 mg/l of oxygen for very polluted streams or raw effluents (Table 1.1.2). This is far in excess of the dissolved oxygen capacity of 7-14 mg/l for natural waters. To determine BOD values then, it is normal procedure to carry out tests on samples diluted to varying extents with uncontaminated water, so that one or more of the mixtures will give a measurable dissolved oxygen consumption of 4-6 mg/l.

TABLE 1.1.2 Typical BOD Values for some Waters and Waste Waters

	BOD (mg/l)
Unpolluted streams	<1
Heavily polluted streams	>10
Secondary treated sewage effluent	5 - 10
Raw sewage and industrial effluents	300 - 5000

A theoretical oxygen demand can be calculated on the basis of the *complete* chemical oxidation of the organic material. Taking the example of glucose (Equation 1.1.2), it can be calculated that 180g of glucose reacts with 192g of oxygen, a ratio of 1:1.07 w/w. Similarly, in the case of sucrose the equation is:

$$C_{12}H_{22}O_{11} + 12O_2 \rightarrow 12CO_2 + 11H_2O + energy \quad \text{(Equation 1.1.3)}$$

Thus 332g of the carbohydrate consumes 384g of oxygen, a ratio of 1:1.16. Even higher specific oxygen uptakes will occur with more highly reduced carbonaceous materials such as starch and fats.

Fig. 1.1.3 Water samples are run from a sampling device used to collect water at depth. Analyses for dissolved oxygen are subsequently carried out in the laboratory.

OBJECTIVES

The main objectives are:

1. to measure the dissolved oxygen content of water samples

2. to measure semi-quantitatively the relative rates of respiration and photosynthesis of some freshwater algae

3. to account for the pH changes in water containing respiring and photosynthesizing algae

4. to calculate the theoretical oxygen demands of some solutions of simple organic compounds

5. to carry out a BOD analysis of some polluted water.

MATERIALS

Aquaria, 5 l
DO/temperature monitors
BOD bottles
incubator
mortar and pestle
measuring cylinders
volumetric flasks
pipettes
magnetic stirrers
spatula
beaker
pH meter
aluminium foil.

Sucrose
glucose
starch
BOD dilution water (dechlorated tap water)
polluted water
pond water containing unicellular algae
plankton net, 1 m^2.

PROCEDURE

The activity should cover two laboratory sessions. The first session is devoted to familiarization with a dissolved oxygen meter, DO measurement, and the preparation of mixtures for BOD analysis.

Using the Dissolved Oxygen Meter

1. Check the condition of the instruments according to the maker's instructions. Take particular care of the gas-permeable membrane at the tip of the probe. This should have no punctures. If the probe has been standing in water for some time, *gently* wipe the membrane surface to remove algal or bacterial slimes.

2. Fill a container with water that has been air-saturated. A BOD bottle is convenient. Measure the water temperature. Insert the DO probe, and stir the water fast enough to obtain a steady reading. Adjust the meter to obtain the appropriate value shown in Table 1.1.3.

3. Reaerate the water and check that the DO reading is the same. If a further check on the probe calibration is needed, the dissolved oxygen content can be assayed titrimetrically using the Winkler method described in most standard texts on water analysis.

Small-Scale Examples of Various Water Quality Conditions

Five small aquaria are set up to illustrate the following water-quality conditions: oligotrophic, eutrophic, anaerobic (rich in biodegradable organic material such as glucose and bacto-agar), saline and heated. Measure the conductance or salinity, temperature, and dissolved oxygen in each case. Express the dissolved oxygen in both mg/l and percent saturation.

Biochemical Oxygen Demand of Prepared Mixtures

1. Weigh out precisely between 80 - 100 mg of glucose and dissolve this in 250 ml of dilution water in a volumetric flask (use dechlorinated, aerated tap water as dilution water). Take 10 ml of the solution and

TABLE 1.1.3 Solubility of Oxygen in Water Exposed to Water-Saturated Air*

% NaCl	0	0.8%	1.6%	2.5%	3.3%
Temp. °C		Dissolved Oxygen (mg/l)			
0	14.6	13.8	13.0	12.1	11.3
1	14.2	13.4	12.6	11.8	11.0
2	13.8	13.1	12.3	11.5	10.8
3	13.5	12.7	12.0	11.2	10.5
4	13.1	12.4	11.7	11.0	10.3
5	12.8	12.1	11.4	10.7	10.0
6	12.5	11.8	11.1	10.5	9.8
7	12.2	11.5	10.9	10.2	9.6
8	11.9	11.2	10.6	10.0	9.4
9	11.6	11.0	10.4	9.8	9.2
10	11.3	10.7	10.1	9.6	9.0
11	11.1	10.5	9.9	9.4	8.8
12	10.8	10.3	9.7	9.2	8.6
13	10.6	10.1	9.5	9.0	8.5
14	10.4	9.9	9.3	8.8	8.3
15	10.2	9.7	9.1	8.6	8.1
16	10.0	9.5	9.0	8.5	8.0
17	9.7	9.3	8.8	8.3	7.8
18	9.5	9.1	8.6	8.2	7.7
19	9.4	8.9	8.5	8.0	7.6
20	9.2	8.7	8.3	7.9	7.4
21	9.0	8.6	8.1	7.7	7.3
22	8.8	8.4	8.0	7.6	7.1
23	8.7	8.3	7.9	7.4	7.0
24	8.5	8.1	7.7	7.3	6.9
25	8.4	8.0	7.6	7.2	6.7
26	8.2	7.8	7.4	7.0	6.6
27	8.1	7.7	7.3	6.9	6.5
28	7.9	7.5	7.1	6.8	6.4
29	7.8	7.4	7.0	6.6	6.3
30	7.6	7.3	6.9	6.5	6.1
31	7.5				
32	7.4				
33	7.3				
34	7.2				
35	7.1				

*at an atmospheric pressure of 760mm Hg

Note: *Use this Table ONLY when it is necessary to calibrate with airsaturated water samples containing known or estimated concentrations of salinity in the ranges listed.*

dilute to one litre; this should form a test solution containing about 3-4 mg/1 of glucose.

2. Prepare test solutions using sucrose and starch in place of glucose. With starch, the weighed sample will have to be ground to a thin paste with a mortar and pestle before making up a suspension in water.

3. Take four thoroughly cleaned BOD bottles (make sure that the bottles have been cleaned with detergent and well rinsed with warm water). Fill three with the test solutions of carbohydrate and fill the fourth with dilution water. Measure the dissolved oxygen in each of the solutions. Top up the BOD bottles if necessary, making sure that no air bubbles are trapped inside. Seal the bottles and incubate at 20°C for one week. The standard period for BOD incubation is 5 days, but here 7 days is suggested to fit more readily laboratory timetabling.

4. After 7 days, remeasure the dissolved oxygen. Calculate the BOD in terms of oxygen consumed using the equation:

$$BOD \text{ (mg/1)} = D_1 - D_2$$

where D_1 and D_2 are the initial and final dissolved oxygen contents respectivel The BOD of the dilution water should be negligible (<0.5 mg/1).

5. Calculate the concentrations of the test solutions of carbohydrate. Calculate the theoretical oxygen demand for each carbohydrate solution as outlined in the Introduction and assuming that starch has the emperical formula $C_6H_{10}O_5$.

Biochemical Oxygen Demand of Polluted Water

Polluted water may be high in oxygen-demanding wastes and thus it may be necessary to dilute before BOD analysis. Several dilutions should be made so that at least one has a 7-day dissolved oxygen drop in the measurable range of 4-6 mg/1.

1. Aerate the polluted water sample. Fill a BOD bottle, measure the dissolved oxygen and incubate at 20°C for a week. Similarly prepare and incubate two other BOD bottles containing polluted water that has been diluted 1:4 (25%) and 1:10 (10%) with dilution water.

2. After incubation, remeasure the dissolved oxygen. Calculate the BOD of the polluted water using the equation:

$$BOD \text{ (mg/1)} = \frac{D_1 - D_2}{P}$$

where P is the dilution factor (1, 0.25 or 0.10). Ignore measurements where $D_2 < 1$ mg/1 (insufficient dilution) or $D_1 - D_2 < 2$ mg/1 (too much dilution).

Effects of Aquatic Plants on the pH and Oxygen Content of Water

1. Obtain about two litres of pond water containing unicellular green algae. Filter off the algae by pouring the water through a piece of plankton net. Transfer the algae to about one litre of BOD dilution water and mix well.

2. Take about 50 ml of the new algae suspension and measure its pH with a meter. Discard this portion.

3. Fill two BOD bottles with the algae suspension. Measure the dissolved oxygen in the contents of one bottle and record this as the initial dissolved oxygen for both. Seal both bottles and light-proof one by wrapping in aluminium foil.

4. Place both bottles in a brightly sunlit position. Place both in a wide shallow tray filled to about 20 cm with water so as to thermostat the bottles.

5. After about 1 hour, remeasure the pH and dissolved oxygen in both algal mixtures. Record the changes that have occurred.

QUESTIONS

1. What are the main causes of the varying dissolved oxygen levels in the five small test aquaria? How is it possible for aquatic plants to cause oxygen "supersaturation" in water?

2. Which process, respiration or photosynthesis, is faster in green plants? Do your results illustrate this?

3. Why does the pH of water containing aquatic plants change depending on whether or not there is light? Use chemical equations in your answer. What are the major forms of dissolved inorganic carbon in water?

4. Comparing the BOD's against the theoretical oxygen demands for glucose, sugar and starch, what can be deduced regarding the biodegradability of these carbohydrates?

BIBLIOGRAPHY

Connell, D.W. (1975), *Water Pollution - Causes and Effects in Australia*, Queensland of University Press, Brisbane.

Higgins, I.J. and Burns, R.J. (1975), *The Chemistry and Microbiology of Pollution*, Academic Press, London.

Odum, E.P. (1971), *Fundamentals of Ecology*, 3rd Ed., Saunders, Philadelphia.

1.2 EUTROPHICATION

INTRODUCTION

The basic building blocks of living systems are plants as these are the only biota which can absorb sunlight energy needed to drive food webs. Sunlight energy is absorbed in the photosynthetic processes to produce green plants. This can be simply expressed as:

$$CO_2 + H_2O + \text{trace elements (N,P etc)} + \text{energy} \rightarrow$$

$$\text{carbohydrate} + \text{protein} + \text{fats (green plants)} + O_2 \quad \text{(Equation 1.2.1)}$$

As a general rule carbon dioxide is readily available from the atmosphere and water from rainfall, ground water or the water mass in aquatic systems. But in most situations the trace elements, particularly N and P, are insufficient to construct the amount of green plant material which would be permitted by the available carbon dioxide and water. Thus N and P are important elements in controlling green plant growth, and usually addition of these elements, in the form of phosphate, nitrate, nitrite and ammonium ions, will stimulate plant growth. For this reason N and P are often described as "plant nutrients."

Lakes, rivers, and most artificial reservoirs are maintained by run-off water from the surrounding land. Contained in this water are plant nutrients, silt and other materials derived from the soil and rocks. Lakes, enclosed bodies of water and, to a lesser extent, slow-flowing streams tend to trap these materials. If the quantities involved are small, the water is usually clear, relatively free of plant growth, and the lake bed is clean sand, gravel or rocks. This condition is generally described as "oligotrophic".

Over geological time, plant nutrients and silt may accumulate. This stimulates the growth of algae and rooted aquatic plants so that fish and other animals may flourish under optimum conditions. This process of enrichment is called "eutrophication". A continuation of this natural ageing leads to the formation of marshes and bogs, and ultimately forests may cover the area.

The activities of man have greatly increased the input of nutrient-rich materials into lakes through such discharges as treated or untreated sewage, crop fertilizers and storm water. The discharge of untreated or primary treated sewage into an aquatic area causes deoxygenation of the water (see Section 1.1) and during this process nutrients are released. Secondary treated sewage has had the Biochemical Oxygen Demand (see Section 1.1.1 and 1.1.3) removed but contains the same nutrients as would be released by untreated sewage. So untreated, primary and secondary treated sewage all contribute nutrients to aquatic areas. The most important nutrients are NH_4^+, NO_3^-, NO_2^-, and $PO_4^{\equiv}$ which occur substantially as a result of the degradation of proteinaceous matter and detergents.

This nutrient enrichment greatly accelerates and intensifies the eutrophication process. The numbers of algae and other green plants often grow uncontrollably to form large dense masses referred to as "blooms". When the plants die and fall to the bottom, their decay occurs through respiration (see Section

1.1.1) and can cause oxygen depletion of the water. When anoxic conditions are reached, hydrogen sulphide and other noxious gases are produced. As well, excessive growths of algae and rooted aquatic plants prevent water movement, and this causes a retardation of dissolved oxygen replenishment. Most aquatic animals cannot survive under these conditions of advanced eutrophication.

Fig. 1.2.1 Eutrophic lake exhibiting a dense, matted growth of aquatic vegetation.

A variety of parameters are used as indices of eutrophication. Tables 1.2.1 and 1.2.2 summarise some of the characteristics which have been used. The phosphorus content of the water is one parameter, as phosphorus is often a limiting nutrient for plant growth. A second is the water's productivity or the rate of photosynthesis, a process that causes the formation of oxygen and the fixation of carbon as described in Equation 1.2.1. Another parameter is the amount of chlorophyll in the water, a measure of the amount of viable algae present.

TABLE 1.2.1 General Characteristics Often Used to Classify Aquatic Areas

Parameter	Oligotrophic	Eutrophic
Aquatic plant production	low	high
Aquatic animal production	low	high
Aquatic plant nutrient flux	low	high
Oxygen in the bottom layers	present	absent
Depth	tend to be deeper	tend to be shallower

(Adapted from Lee (1970), *Eutrophication*, University of Wisconsin Water Resources Centre, Eutrophication Information Program, Occasional Papers No. 2)

TABLE 1.2.2 Guidelines for Classifying Water Bodies According to Primary Productivity, Nutrient and Chlorophyll-a Content

Classification	Primary Productivity ($mgC/m^2/day$)	Total P (mg/l)	Chlorophyll-a ($\mu g/l$)
oligotrophic	0 - 136	<0.01	0.3 - 2.5
mesotrophic		0.01 - 0.03	1 - 15
eutrophic	410 - 547	>0.03	5 - 140

(Loosely based on Vollenweider R.A. (1971), *Scientific Fundamentals of the Eutrophication of Lakes and Flowing Waters with Particular Reference to Nitrogen and Phosphorus as Factors in Eutrophication*, OECD, Paris.)

OBJECTIVES

The main objectives are:

1. to determine the phosphorus content, productivity, and chlorophyll content of a body of water

2. to use the above parameters to assess the trophic status of the body of water.

MATERIALS

Dissolved oxygen meter
water sampler
spectrophotometer
BOD bottles
filter assembly
pipettes
measuring cylinders
volumetric flasks
centrifuge vials
test tubes
centrifuge (optional)
blender (optional)

Ammonium persulphate
18 M sulphuric acid
2 M hydrochloric acid
2 M sodium hydroxide
phenolphthalein
acetone-water (9:1)
combined reagent - this is prepared *fresh* by mixing 50 ml of 2 M sulphuric acid, 5 ml of potassium antimony tartrate hemhydrate solution (1.37g per 500ml), 15 ml of 4% ammonium molybdate and 30 ml of 0.1 M ascorbic acid solution, mixed in the order given.

PROCEDURE

Ideally, a natural body of water should be studied in this activity. If this is not practical, a water tank or large aquarium stocked with an algal culture may be used. The tank should be placed in an open, sunny position, and filled with dechlorinated water. Secondary treated sewage effluent provides a good source of nutrients to encourage algal growth.

Phosphorus

The total phosphorus content is determined by firstly performing an oxidative acid digestion to convert the various forms of phosphorus to orthophosphate ion (PO_4^{3-}). The orthophosphate is then reacted with ammonium molybdate, potassium antimonyl tartrate and ascorbic acid to produce the coloured substance, molybdenum blue.

1. If the water sample contains significant particulate matter, homogenize in a blender for 2-3 minutes.

2. Measure 100 ml into a 250 ml conical flask. Place 100 ml of distilled water in a separate flask - this will serve as a control.

3. Add 1 ml of 18 M sulphuric acid, 0.8 g of ammonium persulphate, and boil gently for 1½ hours. Keep the volume to 25-50 ml with distilled water.

4. Cool, add one drop of phenolphthalein, and neutralize to a *faint* pink colour with 2 M sodium hydroxide solution.

5. Just discharge the pink colour by dropwise adding 2 M hydrochloric

acid. Make up to 100 ml in a volumetric flask with distilled water.

6. For the colourimetric analysis, pipette 20 ml of the sample into a dry test tube, add 1 ml of combined reagent, shake and leave to stand for 10 minutes.

7. Read the absorbance at 880 nm on a spectrophotometer. Use as a reference solution 20 ml of distilled water plus 1 ml of mixed reagent.

8. Prepare several standard phosphorus solutions containing 0.1-1 mg/l P (4.39 g KH_2PO_4 per litre gives 1000 mg/l P).

Treat these as in 6 and 7 above to obtain a standard curve of absorbance *versus* phosphorus concentration.

9. Express the phosphorus concentration of the water as milligram per litre of phosphorus.

Productivity

The production of green plant matter (productivity) occurs as a result of photosynthesis with a proportionate production of oxygen (see generalised Equation 1.2.1). So the determination of oxygen produced in a water body will serve as a measure of productivity. However, respiration also occurs in any water body resulting in a consumption of oxygen so that a measure of this loss is also needed to calculate productivity. The productivity is determined *in situ* by the measurement of the dissolved oxygen change in sub-samples held in "light" and "dark" bottles. In essence, oxygen production measured in the "light" (clear) bottle will provide data on photosynthesis which can be corrected for respiration by using the oxygen reduction in the "dark" (covered) bottle.

1. Obtain four BOD bottles, and light-proof two by wrapping in aluminium foil. These act as the "light" and "dark" bottles, respectively.

2. Obtain a sample of water from the middle of the upper half of the water body. Measure the temperature and dissolved oxygen of the water. If the dissolved oxygen is not supersaturated in the water, record this value as O_i. Fill a pair of light and dark bottles from the remainder of the water sample.

3. If the dissolved oxygen is supersaturated in the water, remove the excess oxygen by gently aerating the water sample. Remeasure the dissolved oxygen and record as O_i. Fill a pair of light and dark bottles with the remainder of the water sample.

4. Obtain a sample of water from the middle of the bottom half of the water body and treat this as in 2 or 3 above.

5. Suspend the two pairs of light and dark bottles in the water body at depths corresponding to where the water samples were taken. Arrange the bottles to avoid shading from the sunlight. Fig. 1.2.2 shows a convenient way of arranging the bottles.

6. Expose the bottles for several hours. The usual exposure period is from dawn to noon or noon to dusk, or for the whole photoperiod. A shorter time may be chosen for convenience.

7. At the end of the exposure period, remove the bottles and measure the dissolved oxygen in each. Record the values as O_l and O_d for the light and dark bottles respectively.

8. Determine the oxygen changes due to respiration as well as net and gross photosynthesis using the formulae:

Respiration, R = O_2 decrease in dark bottle = $O_i - O_d$

Net Photosynthesis, Pn = O_2 increase in light bottle = $O_l - O_i$

Gross Photosynthesis, Pg = Respiration + Net Photosynthesis

$$= (O_i - O_d) + (O_l - O_i)$$

$$= O_l - O_d$$

9. Average the two sets of data for the upper and lower halves of the water body.

10. Estimate the daily rate of gross and net phototsynthetic activity per unit area of water by carrying out the following calculations:

(i) Correct the exposure time to a daily period:

Daily Pg (mg O_2/l/day) = Pg x $\frac{\text{length of photoperiod per day}}{\text{length of exposure time}}$

(ii) Change the productivity units from mg/l of oxygen to mg of oxygen per square metre of water surface. This expresses the total productivity in the water column under 1 m^2 of water surface. The depth of water in the productive zone must be known.

Daily Pg (mg O_2/m^2/day) = Pg x $\frac{\text{length of photoperiod per day}}{\text{length of exposure time}}$

x 10^3 x water depth (m)

The factor 10^3 converts the volume concentration from mg/l to mg/m^3.

(iii) Calculate the Daily Respiration assuming that the respiration remains constant for a whole 24 hour day:

Daily R (mg O_2/m^2/day) = R x $\frac{24}{\text{exposure time (hr)}}$ x 10^3 x water depth (m)

(iv) Calculate the Daily Net Photosynthesis:

Daily Pn (mg O_2/m^2/day) = Daily Pg - Daily R.

11. Convert the productivity from units of oxygen produced to units of carbon fixed assuming that the idealized equation for phtotsynthesis (CO_2 +

$H_2O \rightarrow CH_2O + O_2$) holds:

$$\text{Daily Pn (mg C/m}^2\text{/day)} = \text{Daily Pn (mg O}_2\text{/m}^2\text{/day)} \times \frac{12}{32}.$$

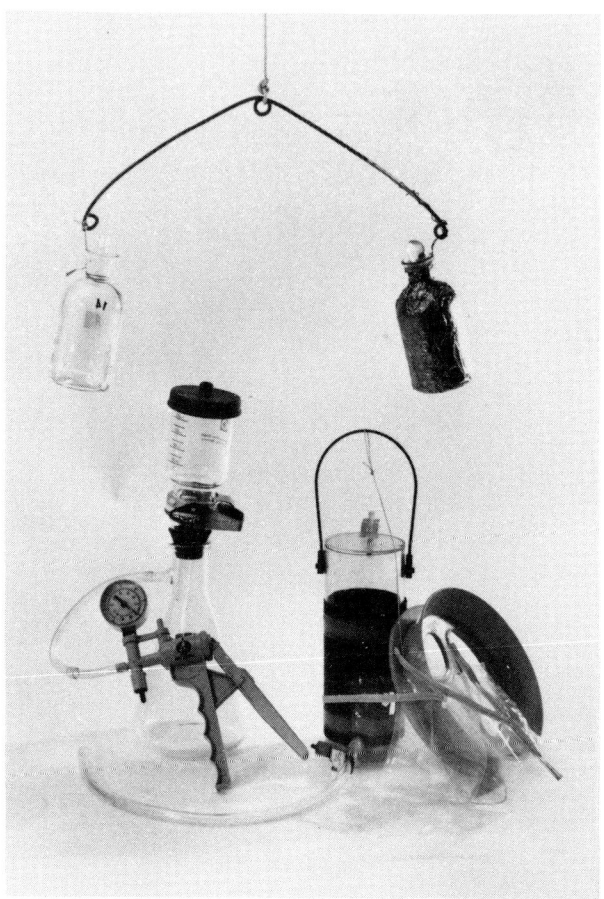

Fig. 1.2.2 Some field equipment used in the assessment of eutrophication: a cylinderical water sampler equiped with trapdoors and cord; filter assembly and pump; a pair of light and dark bottles, several of which may be connected in series; plastic sample bags. (Photo: John Watson)

Chlorophyll-a

An estimate of the green plant matter in a water body can be obtained by measuring the chlorophyll-a content of the water body. The chlorophyll-a content is determined by extraction of the pigment into acetone followed by the measurement of its visible absorbance.

1. Filter 100-500 ml of the water sample through a glass-fibre filter. Record the volume filtered.

2. Roll the filter into a cigarette shape, and place it in a small vial or centrifuge tube.

3. Add 10 ml of 90% acetone, or enough to cover the filter. Record the volume. Stopper the vial, and store in the dark at $4^{\circ}C$ for 24 hours.

4. Centrifuge the extract if it is cloudy.

5. Transfer some of the extract to a 1 cm glass cuvette. Measure the absorbance at 665 and 750 nm. Subtract the 750 nm value from the 665 nm value. The 750 nm value serves to correct for any turbidity.

6. Add one drop of 2 M hydrochloric acid to the cuvette. Mix, and stand for 1 minute. Remeasure the absorbance at 665 and 750 nm and subtract the values. The remaining figure is taken as the absorbance due to pigments other than chlorophyll-a.

7. Calculate the chlorophyll-a concentration using the formula:

$$\text{Chl-}a \ (\mu g/l) = 29(A-A_a) \times \frac{\text{vol. extract (ml)}}{\text{vol. sample (l)}}$$

where A is the corrected absorbance at 665 nm before acidification and A_a is the corrected absorbance at 665 nm after acidification.

QUESTIONS

1. What is the trophic status of the water body according to the results for (a) phosphorus, (b) productivity, and (c) chlorophyll-a?

2. Why is phosphorus a common limiting nutrient? What are the main sources of phosphorus in aquatic areas?

3. What are some other main limiting nutrients apart from phosphorus, in order of importance? What are their major sources?

4. The estimation of daily productivity involved several major assumptions. What are some of these?

BIBLIOGRAPHY

Hart, B.T. (1974), *A compilation of Australian Water Quality Criteria*, Australian. Water Resources Council Technical Paper No. 7, Australian Government Publishing Service, Canberra.

U.S. Environmental Protection Agency (1976), *Methods for Chemical Analysis of Water and wastes*, USEPA Technology Transfer EPA-625-/6-74-003a, Washington.

Vollenweider, R.A. (Ed.) (1974), *A Manual on Methods for Measuring Primary Production in Aquatic Environments*, IBP Handbook No. 12. Blackwell Scientific, London.

1.3 SEWAGE TREATMENT - A FIELD TRIP

INTRODUCTION

The major pollutant produced by most societies is domestic sewage. This consists of liquid wastes derived from homes, business operations and many industries which are collected in a sewerage system and transferred to a treatment plant. It usually contains faecal matter, food processing wastes, sullage and a variety of other wastes referred to as organic wastes.

The major chemical components of sewage are carbohydrates, proteins and fats and a variety of chemical combinations of these substances (see Table 1.3.1). In addition, microbiological activity is high in domestic sewage.

The degradation of sewage occurs by basically the same biochemical processes whether in a treatment plant or in a natural aquatic area. In simple terms the patterns of degradations can be outlined as in Fig. 1.3.1. The simplest case is the aerobic degradation of glucose by the respiratory processes of microorganisms and can be represented overall by the following equation:

$$C_6H_{12}O_6 + 6O_2 \rightarrow 6CO_2 + 6H_2O$$

TABLE 1.3.1 Composition of Fresh Sewage Sludge

Component	Approximate Percentage
Ether soluble (oil, fats)	34.4
Cold and hot water soluble (aminoacids, starch, pectin acids)	9.5
Alcohol soluble	2.5
Hemicellulose	3.2
Cellulose	3.8
Lignin	5.8
Crude protein	27.1
Ash	24.1
Approximate total	110*

(From: Dugan, P.R. (1975), *Biochemical Ecology of Water Pollution*, Plenem Press, New York, p.70)

---

* This total is more than 100% since several components overlap in composition.

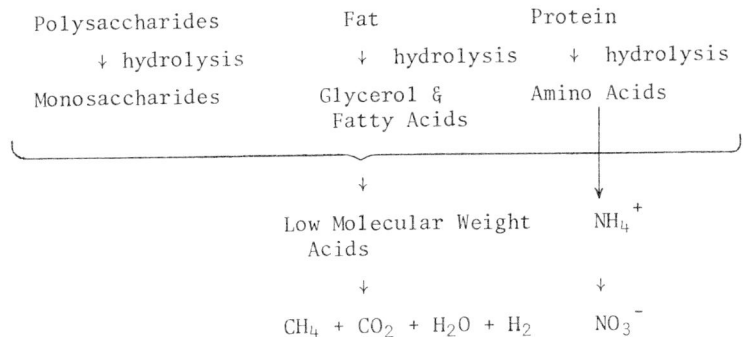

Fig. 1.3.1 Degradation patterns of the principal components of sewage.

Under the same conditions the N content of proteins would be substantially converted into nitrate salt (NO_3^-) and phosphorus into orthophosphate ($PO_4^{\equiv}$). Under anaerobic conditions the degradation of glucose can be ideally expressed as:

$$12CH_2O \rightarrow 6CH_4 + 6CO_2$$

SEWAGE TREATMENT TECHNIQUES

The pollution of rivers by sewage dates from the establishment of the earliest towns and cities. In 1653, Izaak Walton published the *Compleat Angler* and made no mention of pollution although other records indicate that there were a few relatively isolated examples of river pollution. However, by the 1800's large urban complexes had developed in which sewage was discharged directly into rivers untreated. The introduction of new industries and sewers which served to transport wastes more efficiently to streams aggravated the stream pollution problem while generally improving public health.

Nevertheless water-borne diseases, such as cholera and typhoid were serious problems causing large numbers of deaths. In Britain a Royal Commission was set up in 1857 and was followed by another in 1865. These activities resulted in the passage through Parliament of the Rivers Pollution Acts of 1876 and 1890, and a recommendation for the treatment of sewage by "sewage farms", in other words disposal on farm lands. A number of large sewage farms were set up in various parts of the world and some are still operational although more efficient methods are now available.

The objectives of wastewater treatment fall principally in three classes:

1. To remove the B.O.D. by biological degradation of the organic substances by microorganisms. This process is carried out by aerobic or anaerobic respiration.

2. To discharge effluents which do not contain microorganisms harmful to human health.

3. To remove substances which have an undesirable impact on the environment such as plant nutrients.

The objectives in any specific case are influenced by such factors as:

>   Economic resources available
>   The particular environment into which discharges are made
>   Possible reuse of the effluent water
>   Uses which are made of the aquatic environment into which discharges are made.

To meet the objectives and to relate them to the factors outlined above, sewage can be treated to a variety of degrees of purification. As a general rule the treatment procedures can be divided into 3 main classes depending on the degree of purification achieved.

1. Primary Treatment - Removal of solids and sludge prior to discharge.

2. Secondary Treatment - Primary treatment plus aeration and treatment by the activated sludge technique to remove the substances giving B.O.D. Other processes based on aeration can be used for secondary treatment also.

3. Tertiary Treatment - Secondary treatment plus removal of inorganic nitrogen and phosphorus salts which act as plant nutrients.

A general description of the treatment techniques used today is outlined below:

Sewage Farm/Land Disposal

This method usually consists of a preliminary screening and discharge of the liquid waste onto farm land. Aerobic degradation occurs, but run-off water into adjacent waterways can be comparatively high in B.O.D. Agricultural products produced in this way may be required to pass certain public health tests.

Lagoons, Stabilisation or Oxidation Ponds

These are simply large ponds usually about 1-2 m deep where sewage is discharged and held for periods up to several months. Aerobic degradation occurs as a result of exposure to air and, in addition, sunlight stimulates the growth of algae which help provide oxygen for the respiration process. This can be an effective treatment process depending on residence times of the sewage and weather conditions.

Primary Treatment

This process consists of screening of the sewage and removal of sludge in a sedimentation tank. The liquid discharge is usually high in B.O.D. but populations of microorganisms can be destroyed by chlorination (see Fig. 1.3.2).

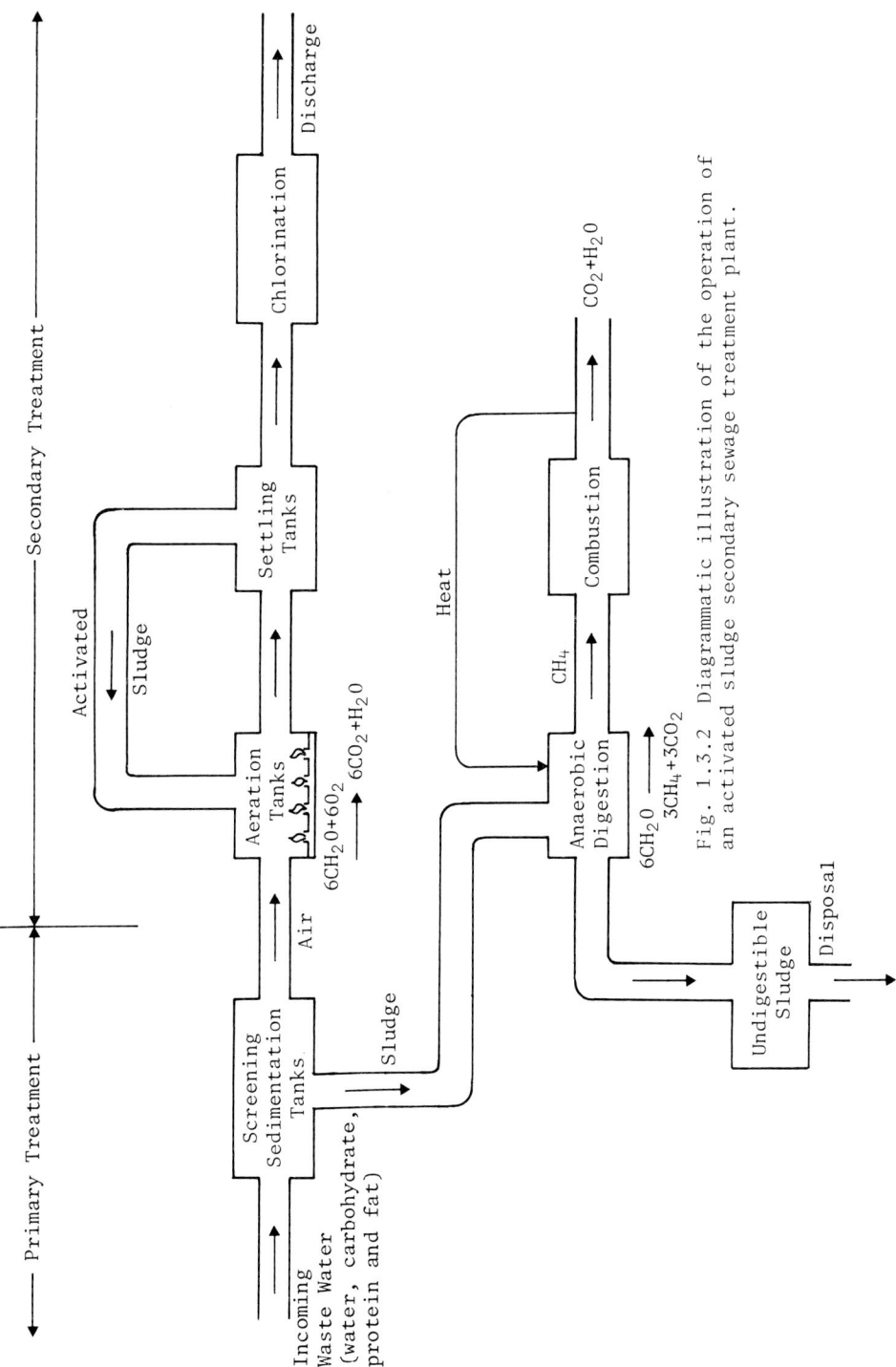

Fig. 1.3.2 Diagrammatic illustration of the operation of an activated sludge secondary sewage treatment plant.

Fig. 1.3.3 A modern secondary sewage treatment plant
utilising (1) sedimentation tanks (2) anaerobic digesters
(3) electricity generation plant using methane from the
anaerobic digesters (4) biological filters (5) oxidation
ponds (6) laboratory (Photo: South Australian Government).

Secondary Treatment by Activated Sludge

An outline of primary treatment followed by secondary treatment is shown in
Fig. 1.3.2. In this method the B.O.D. and solids contents of the sewage
are reduced to low levels, usually the B.O.D. is less than 10 mg/l and the
suspended solids less than 20 mg/l. The undigestable sludge produced is
usually disposed of in land fill. It is noteworthy that the methane produced
in this process can be used to maintain the anaerobic digestor at an optimum
temperature, usually above atmospheric temperatures. In many plants the
methane is also used in electricity generation to run pumps, etc, in the
plant.

Septic Tanks

A septic tank can be attached to individual homes which are not served by
sewerage systems. This consists of an underground tank which accepts
sewage and aerobic degradation then occurs. The purified wastewater is

discharged into the ground through sub-surface drains. After a period of operation the accumulated sludge in the tank should be removed. The efficiency of this process is related to the capacity of the ground to accept wastewater.

Secondary Treatment by Trickling or Biological Filters

With this method the sewage is primary treated then pumped to a filter. The filter consists of a bed of stones usually 1-3 m deep onto which the primary treated sewage is discharged at a slow rate by a rotating arm or other method. Bacteria and a variety of other organisms grow on the bed and aerobically degrade the sewage as it passes through. (See Fig. 1.3.3)

Tertiary or Advanced Wastewater Treatment

All of the treatment methods described above result in the production of waste-water containing nitrogen and phosphorus salts. These substances may stimulate excess plant growth and cause eutrophication (see Section 1.2). In many situa-tions there is a need to remove these salts and produce a very high quality effluent. A wide range of techniques can be used ranging from extensions of the biological methods to physico-chemical methods such as adsorption, distillation and osmosis.

OBJECTIVES

The main objectives are:

1. to observe the sewage treatment methods used

2. to assess the chemical changes used in sewage treatment

3. to observe the end products of sewage treatment

4. to consider the environmental effects of sewage treatment.

QUESTIONS

In writing up this activity there is no need to give a detailed description of the plant or its operation. However, the following questions should be answered:

1. Write idealised chemical equations for the degradation of carbohydrates occurring in various parts of the plant e.g.

    (a) the activated sludge - aeration tanks
    (b) the sludge digestion tanks

2. What happens to the methane produced in the plant?

3. Is this plant a primary, secondary or tertiary treatment plant? What are the basic treatment processes carried out in the plant?

4. What types of substances occur in the effluent from the plant and what is their possible environmental impact?

5. What possible uses could be made of the effluent from the plant?

6. What kind of substances are in the undigestible sludge?

7. What possible uses could be made of the undigestible sludge usually produced as a dried solid? Are there difficulties in its disposal?

8. What important substances and processes are monitored during the plant's operations?

BIBLIOGRAPHY

Dugan, P.R. (1975), *Biochemical Ecology of Water Pollution*, Plenum Press, New York.

Higgins, I.J. and Burns, R.G. (1975), *The Chemistry and Microbiology of Pollution*, Academic Press, London

PART 2

Toxic Substances in the Environment

2.1 INSECTICIDES IN CIGARETTE SMOKE

INTRODUCTION

During the three decades since the second world war, millions of tonnes of pesticides have been distributed in the environment to control plant and insect pests. Residues of these pesticides or their breakdown products occur worldwide, contaminating air, water, soil, plants, animals and man himself.

The main groups of pesticides are insecticides, herbicides and fungicides with minor groups of rodenticides, avicides, molluscicides, acaricides, nematocides, bactericides and antivirals. Taking the insecticides, these are classified into subgroups, according to their structure and function as outlined below.

The Organochlorine or Chlorinated Hydrocarbon Insecticides

Examples are DDT, dieldrin, aldrin, DDD, heptachlor, lindane, and DDE:

DDT

DDD

Aldrin

Lindane

Dieldrin

Heptachlor

DDE

Organophosphorus Insecticides

Organophosphorus insecticides are generally derivatives of phosphoric acid or the sulphur analogues of phosphoric acid:

$$R_1-O, R_2-O \!\!>\!\! P(=O \text{ or } S)-O(\text{or } S)-R_3$$

The groups R_1, R_2 are usually the same, most often the alkyl groups CH_3 or CH_3CH_2. The type of R_3 group is much more varied, and includes alkyl, aryl and heterocyclic groups and their derivatives. The variety of chemical types which can be used as this group accounts for most of the diversity in organophosphorus insecticides.

Organophosphorus insecticides often encountered are malathion, a common garden insecticide, and dichlorvos, a volatile compound that is the active ingredient of "pest strips":

$$CH_3O, CH_3O \!\!>\!\! P(=S)-SCHCO_2CH_2CH_3$$
$$|$$
$$CH_2CO_2CH_2CH_3$$

Malathion

$$CH_3O, CH_3O \!\!>\!\! P(=O)-OCH_2=CCL_2$$

Dichlorvos

Carbamate Insecticides

Carbamate insecticides are based on the chemical group:

$$R-O-\overset{O}{\underset{\|}{C}}-N\!\!<$$

Carbaryl is a well known example:

$$O-\overset{O}{\underset{\|}{C}}-N\!\!<\!\!\overset{H}{\underset{CH_3}{}}$$

Carbaryl

Inorganic Insecticides

These are usually compounds of arsenic or mercury. Lead arsenate, $PbHAsO_4$, is an example.

The use of organochlorine insecticides, particularly DDT, has declined in recent years due principally to long-term deleterious effects. Residues of organochlorine insectidices are very persistent, and are accumulated in the fatty tissues of predatory animals. Deaths and the lack of breeding success in predators such as hawks and falcons have been attributed to the accummulation of these residues, and there is a possible risk to man. Nowadays, shorter-lived pesticides are more commonly used, such as organophosphosus compounds and, to a lesser extent, organocarbamate compounds. These alternative pesticides are far less persistent, although they often have higher mammalian toxicity.

The analysis of pesticide residues in environmental samples almost invariably involves some form of chromatographic separation. Thin layer chromotography, or TLC, is a simple and rapid means of separation used where qualitative identification is needed. In this case a glass plate is usually coated with an absorbent such as finely powdered silica gel so that a thin layer is formed. The insecticide mixture is absorbed onto the silica gel as a spot. The plate is stood with the lower portion immersed in a bath of solvent which moves then through the layer by capillary action. The insecticide components move at different rates depending on their affinity for the absorbent. After drying, the new position of the insecticides can be viewed either under UV light or by spraying with certain chromogenic (colour-forming) reagents.

A quantitive value can be placed on the movement of a separated compound on a TLC plate. This is termed the R_f value, and is defined as:

$$R_f = \frac{\text{distance the spot travels from the origin}}{\text{distance the solvent front travels from the origin}}$$

The R_f value is characteristic for a particular compound, and is influenced by chromatographic conditions such as the nature, of the absorbent powder thickness and particle size, the mobile solvent, and the temperature.

OBJECTIVES

The main objectives are:

1. to extract pesticides from cigarette smoke

2. to separate and identify some of the pesticide components using column and thin layer chromatography.

MATERIALS

Side-arm filter tubes
aluminium foil
pasteur pipettes

Cigarettes
dichloromethane
hexane

disposable glass vials
fluorescent TLC plates,
  silicagel GF_{254}, 5x20cm
glass capillary tubing
developing tanks
U.V. viewer
chromatography atomizers

acetone
anhydrous sodium sulphate
Florisil, 60-100 mesh, 5%
  deactivated
glass wool
0.5% o-toluidine in
  ethanol
2% 4-(nitrobenzyl) pyridine
  in acetone
10% tetraethylenepentaamine
  in acetone
standard solutions of DDT,
  DDE, and dichlorvos in
  hexane, 100 µg/ml

PROCEDURE

A cigarette is "smoked" in such a way that the smoke is passed through a solvent which traps much of the volatile tars and pesticide residues. The residues are "cleaned-up" or separated from interfering substances such as fats, tars and colouring material by smale-scale column chromatography prior to TLC analysis.

1. Place about 10ml of dichloromethane in a side-arm filter tube. Join a cigarette to a pasteur pipette and fit in the filter tube as shown in Fig. 2.1.1.

2. Use a mild vacuum to draw air at a gentle rate through the cigarette. Light the cigarette and allow it to burn down.

3. When the "smoking" is complete, evaporate off the dichloromethane using a warm water bath and a stream of dry air. A pale brown, sticky residue will remain.

4. Make up a small chromatography column using a pasteur pipette. Push a small plug of glass wool to the narrow section of the pipette, then add Florisil to form a column about 5 cm long. Top the Florisil with about 1 cm of anhydrous sodium sulphate granules. At least 1 cm of the top of the pastuer pipette should remain empty.

5. Rinse the column with about 5 ml of hexane and discard the waste. Dissolve the sticky residue in about 1 ml of hexane and transfer this to the top of the column.

6. Elute the column with about 4 ml of hexane and collect the effluent in a small glass vial. Further elute the column with 10% acetone in hexane, collecting the new effluent in a second glass vial. Evaporate off the solvents in a stream of air with warming. The two glass vials contain residues or organochlorines and organophosphates, respectively.

The insecticide residues are chromatographed on silica-gel thin layer plates using weakly polar solvents as the mobile phase. DDT, DDE and dichlorvos are used as reference standards, but other insecticides may be used if they are more appropriate for the samples chosen.

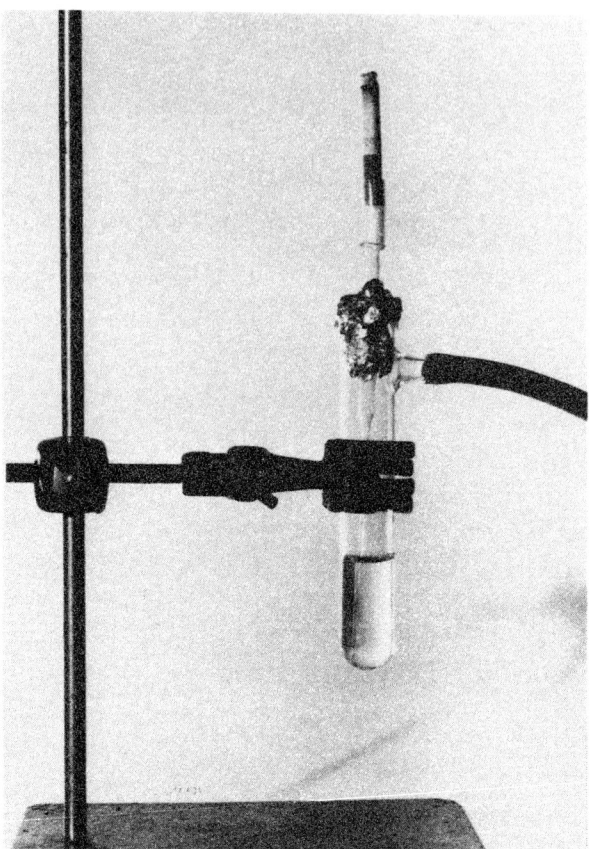

Fig. 2.1.1 Sidearm test tube, Pastuer pipette, plastic tube connector, aluminium foil plug and dichloromethane used for collecting pesticide residues in cigarette smoke. (Photo: John Watson).

7. Trace the outline of a TLC plate on a blank page. Mark on the page a row of three dots about 3 cm from the bottom edge.

8. Add 1-2 drops of hexane to the first vial containing the organochlorine residues. Take this solution up into a short length of glass capillary tubing. With a quick movement, touch the tip of the capillary onto the TLC plate in the spot corresponding to the first dot on your diagram. Blow on the small stain to dry. Repeat this touch-and-blow procedure till all the residue solution is applied.

9. Use a fresh piece of capillary to spot DDT standard on the place corresponding to the second dot on your diagram. Similarly spot DDE standard in the place corresponding to the third dot.

10. Place the TLC plate in a tank containing hexane to a depth of 1 cm. Let the solvent run for about 15 cm. Remove the plate, quickly mark the position of the solvent front, and dry in air.

11. View the place under short-wave u.v. light. Dark, non-fluorescent patches will indicate the new positions of the DDT and DDE spots. The organochlorine spots can be shown more permanently by spraying with o-toluidine solution then u.v. irradiating for about 10 minutes when brown spots will appear. Copy the pattern of spots onto your diagram.

12. In a similar way, spot the organophosphorus residues onto a fresh TLC plate. Spot dichlorvos standard alongside. Develop the plate in a tank containing hexane - acetone (4:1). Dry the plate at 100°C for 2 minutes, then spray with 4-(4-nitrobenzyl)-pyridine solution. Redry, then over-spray with 10% tetraethylenepentaamine in acetone. Let the plate stand for about 30 minutes in air, when the organophosphates will appear as blue spots against a white background.

QUESTIONS

1. What are the R_f values of the insecticides tested? Use the chromatographic results to place the insecticides in an order of polarity.

2. What insecticides were identified?

3. DDT was once widely applied to tobacco crops. What are some of the pesticides in use today, both in the field and in the curing and storage?

4. What are the LD_{50} (rat) and the acceptable daily intake (humans) for dichlorvos?

BIBLIOGRAPHY

Fishbein, L. (1975), *Chromatography of Environmental Hazards;* Elsevier, Amsterdam.

Green, M.B., Hartley, G.S., and West, T.F. (1977), *Chemicals for crop Protection and Pest Control,* Permagon Press, Oxford.

Joint FAO/WHO Food Standards Programme Codex Alimentarus Commission (1976), *Recommended Maximum International Limits for Pesticide Residues,* Fifth series, FAO, Rome.

Vettorazzi, G. and Miles-Vetorazzi, P. (1975), *Safety Evaluation of Chemicals in Food: Toxicological Data Profiles for Pesticides. 1. Carbamate and Organophosphorus Insecticides used in Agriculture and Public Health,* WHO, Geneva.

2.2 GAS CHROMATOGRAPHIC INVESTIGATION OF VOLATILE HYDROCARBONS

INTRODUCTION

Of all the waterborne pollutants petroleum oils are the most obvious. Oil slicks and tarry residues are now common throughout the world's waterways. There is little doubt that the bulk of these pollutants have originated from the extraction use and transport of petroleum products. Table 2.2.1 gives a summary of the estimates of quantities and sources of petroleum discharges to the oceans.

TABLE 2.2.1 A Recent Estimate of Petroleum Hydrocarbons Entering the Oceans

SOURCE	MILLIONS OF TONNES/YEAR
Marine transportation and off-shore oil production	2.41
Industrial and municipal waste	0.6
Land run-off	1.9
Natural seeps	0.6
Atmospheric rain-out	0.6
TOTAL	6.11

(From: Joint Group of Experts on the Scientific Aspects of Marine Pollution (GESAMP) (1977), *Impact of Oil on the Marine Environment*, Rep. Stud. No. 6. Food and Agriculture Organisation, Rome).

Petroleum Hydrocarbons

The refining and cracking of petroleum, and to a much lesser extent the destructive distillation of coal, serve as the world's major sources of hydrocarbon products. As well as being of importance as fuels, these substances form the basic components for the synthesis of such materials as synthetic rubber, plastics and many pharmaceuticals. Fuels, such as automobile spirit, kerosene, diesel and fuel oil, as well as lubricating oils, ae among the most common hydrocarbon substances discharged into aquatic areas.

Hydrocarbons can be simply classified into three main groups according their structure:

(a) The Alkanes or "Saturated" Hydrocarbons. These substances are usually the major components of petroleum. Examples are hexane and 2,2-dimethyl-propane:

CH₃-CH₂-CH₂-CH₂-CH₂-CH₃

$$CH_3-\underset{\underset{CH_3}{|}}{\overset{\overset{CH_3}{|}}{C}}-CH_3$$

      n-hexane　　　　　　　　　　　　　　2,2-dimethylpropane

Because of the strength of the C-C and C-H bonds, the alkanes are relatively resistant to degradation by most chemical and biological agents. The alkanes exhibit a very low level of toxicity towards biota. Solubility in water is of a very low order due to the low polarity of the molecule.

(b) <u>The Alkenes and Alkynes or "Unsaturated" Hydrocarbons</u>. The term unsaturated arises from the lower H:C ratios than those for the alkanes, which is caused by the presence of double and triple bonds carbon-carbon. Examples are propene and ethyne:

      CH₃-CH=CH₂　　　　　　　　　　　　　　H-C≡C-H

      propene　　　　　　　　　　　　　　　　ethyne (acetylene)

Multiple bonds are susceptible to disruption, and accordingly the unsaturated hydrocarbons are more easily degraded by environmental agents such as atmospheric oxygen, ultravoilet light and biotic metabolism than are the alkanes. The toxicity and water solubility are generally higher for the alkenes and alkynes than for the alkanes.

(c) <u>The Arenes or Aromatic Hydrocarbons</u>. These hydrocarbons contain one or more six-membered chains of carbon atoms. They are unsaturated in that they have H:C ratios lower than those for the alkanes. Unlike the alkenes and alkynes, however, the additional C-C bonds form as a continuous link connecting all the carbon atoms in the chain. Two examples are:

　　　　　　　　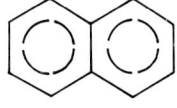

   Benzene (C₆H₆)　　　　　　　　Naphthalene (C₁₀H₈)

This group exhibits markedly different properties to the other two. Some members are moderately soluble in water, most are toxic and several members are highly carcinogenic. Although common constituents of petroleum they are almost invariably in lower proportion than the other two groups.

<u>Investigation of Petroleum Hydrocarbons Using Gas Liquid Chromatography</u>

Petroleum hydrocarbons can occur in water, sediments and aquatic biota. Since they are usually of low water solubility and high fat solubility, concentrations in water are very low whereas the fatty tissues of animals and plants can contain relatively high concentrations.

The most widely used technique for analysing petroleum hydrocarbons is gas liquid chromatography commonly described as gas chromatography. Gas chromatography is a technique for separating mixtures of volatile substances into individual components. It depends on the individual components moving at different speeds when carried through a packed tube or column by an inert gas. The column is usually 1-3m long and several millimetres in diameter, packed with an inert porous solid such as diatomaceous earth, and coated with a nonvolatile liquid. Some liquid coatings in use are silicone rubbers, greases, waxes, oils, carbowaxes, and dialkyl phthalates.

The nonvolatile liquid plays a major role in the separation process. The speed that individual substances move through the column depends principally on the partition of the substance between the gas and the liquid phases. Substances where the partition favours the liquid phase will be retained for longer times in the column than those where the gas phase is favoured.

Fig. 2.2.1 shows diagramatically the design of a gas chromatograph. It consists essentially of the column, usually made of metal or glass tubing, encased in a heating oven. At one end is the injection port and at the other a detector, both also heated. The output from the detector is usually fed to a chart recorder.

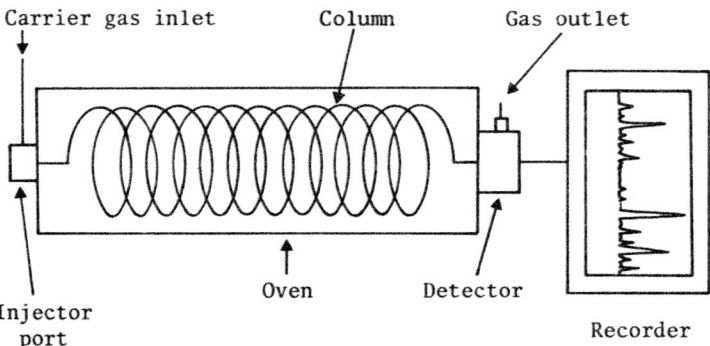

Fig. 2.2.1 A diagrammatic illustration of a gas chromatograph.

In practice, a small volume, about 1 µl, of the mixture or a solution of the mixture is injected into the gas chromatograph. This is immediately volatilized and the vapours swept along the column by the carrier gas, usually an inert gas such as helium, nitrogen or argon. After separation on the column, the components then pass through a detector sensitive to the effluent substances. One such detector is the thermal conductivity detector, a·device which senses the difference in thermal conductivity between the carrier gas and the carrier gas plus a separated component. Thus, as each component passes, the detector emits a signal which is transmitted to the recorder. The recorder trace displays a series of peaks, each peak corresponding to a different compound which has passed through the detector (see Fig. 2.2.1). The area of the peaks can be used to estimate the quantitative composition of the mixture. The time required for a compound to pass through a column is known as its retention time (t_r), which under constant conditions is always the same and is characteristic of the compound. However the retention time does vary under different conditions including such factors as:

(a) the nature of the liquid phase
(b) the concentration of the liquid phase on the solid support
(c) the column temperature
(d) the flow rate of carrier gas
(e) the type of carrier gas
(f) the dimensions of the column.

By suitably adjusting these variables it is usually possible to optimise the efficiency of separation of a mixture. Fig. 2.2.2 shows an example of different separation efficiencies of two substances. Examination of the figure indicates that the width of the peaks recorded is a significant factor in efficiency. Column efficiency is usually measured in terms of *theoretical plates* which derives from the use of distillation columns to separate mixtures. In distillation columns the larger the number of plates the greater the efficiency of separation. In a gas chromatography column, the efficiency can be calculated using the following expression:

$$N = \frac{16(t_r - t_a)^2}{W}$$

where N - number of plates

t_r - retention time

t_a - time for a substance (such as air) which does not interact with the column components to move through the column

W - peak width at the base

Gas chromatography separates chemically similar compounds according to their volativity. Thus, in a homologous series such as the aliphatic hydrocarbons hexane, heptane, octane and nonane, the hydrocarbons will exhibit a regular increase in retention time in the carbon number sequence of C_6, C_7, C_8, and C_9. In practice, a direct relationship holds between the carbon number (or molecular weight) of the hydrocarbon and the logarithm of its retention time, as shown in Fig. 2.2.3. This sort of relationship is important as it allows interpolation or extrapolation to obtain retention times of members of the homologous series which may not be available.

OBJECTIVES

The main objectives are:

1. to operate a simple gas chromatograph

2. to determine the retention times of some pure hydrocarbons

3. to separate and identify the components of a mixture of volatile hydrocarbons.

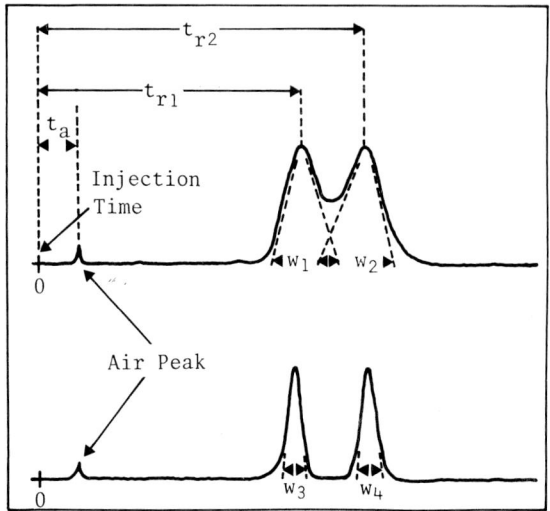

Fig. 2.2.2 Hypothetical gas/chromotograms of two compounds under different conditions of column efficiency.

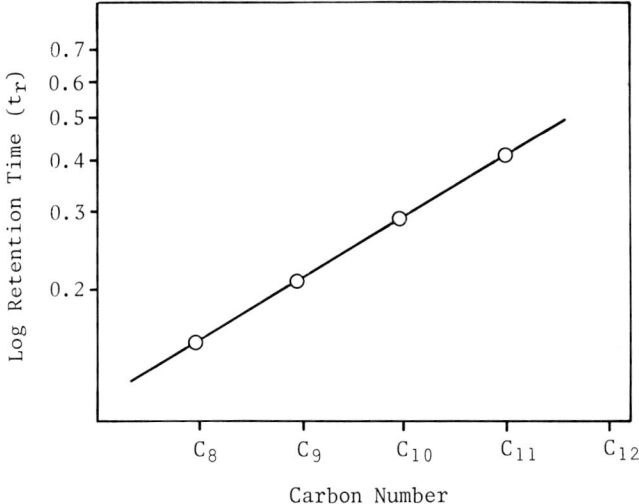

Fig. 2.2.3 Plot of retention time of hydrocarbon *versus* carbon number for the alkanes.

MATERIALS

Gas chromotograph (thermal conductivity detector with liquid phase such as SE30 or Apiezon M)
plastic rulers
1 or 5 µl syringe

Hydrocarbon standards of pentane, hexane, heptane, octane, decane, cyclohexane and isooctane
mixtures of volatile hydrocarbons

PROCEDURE

1. Set up a gas chromatograph according to the following conditions:

    injection, column and detector temperatures at 120-150°C
    helium flow rate at 30ml/min using a soap bubble meter
    recorder at 1 mV full scale, chart drive at about 3 cm/min.

2. Adjust the GC attenuator to its most sensitive setting. A flat pen trace will indicate that the temperatures and gas flows are stabilized.

3. Inject 1 µl of hexane and quickly mark in the injection point on the recorder trace. If the recorder peak goes off scale, rechromatograph 1 µl of hexane using a higher attenuation setting.

4. Chromatograph 1 µl volumes of pentane, heptane, octane, decane, cyclohexane and iso-octane. Note down the necessary details on the recorder paper. Calculate the retention times of each hydrocarbon using the recorder speed and the distance of the peak maxima from the injection points.

5. Plot the carbon numbers of the hydrocarbons against the logarithm of their retention times. Fit the points to a straight line.

An unknown mixture of hydrocarbons is chromatographed to identify its components. Petrol, or a combination of the standard hydrocarbons, makes a useful mixture.

6. Chromatograph 1 µl of the hydrocarbon mixture. Identify the components by matching the peak positions after the injection point with those of the hydrocarbon standards. Not all peaks in the mixture may be able to be matched with the limited numbers of standard hydrocarbons used.

7. Calculate the concentrations of the identified components using either relative peak area or relative peak height. For a given hydrocarbon, the recorder peak area is taken as proportional to the amount of hydrocarbon injected, so that the amount in a mixture is given by;

$$\text{Hc in mixture (µl or µg)} = \text{Hc in standard (µl or µg)} \times \frac{\text{peak area in mixture}}{\text{peak area in standard}}$$

Suitable corrections have to be made if the injection volumes or attenuator settings are changed. The area under a peak maybe obtained by assuming a triangular shape and taking the product of peak height x peak width at half height.

QUESTIONS

1. What is the composition of the hydrocarbon mixtures examined?

2. What are the boiling points of the hydrocarbon standards used? Is there any connection between the boiling points and the retention times of the hydrocarbons?

3. What is the efficiency of the gas chromatography column used?

BIBLIOGRAPHY

Ambrose, D. (1971), *Gas Chromatography*, Butterworth, London.

Grant, D.W. (1971), *Gas liquid Chromatography*, Van Nostrand Reinhold, London.

2.3 TOXICITY OF COPPER IONS TO AQUATIC ORGANISMS

INTRODUCTION

A variety of toxic substances are discharged into the environment where they may exert harmful effects on organisms. Environmental toxicology is broadly concerned with the effects of air and water pollutants, residues in food and biota and industrial hygiene.

Toxic substances have two important aspects to their action. Firstly the physiological impact of the substance on organisms and secondly the concentration in the environment, or dose administered to the organism, which causes that effect. This concept includes the basic principle that harmful effects are related to dose or concentration and below a certain minimum there will be no harmful effects.

These principles are illustrated by the hypothetical data plotted in Fig. 2.3.1. This indicates the lethal response of a uniform population of an organism when subject to increasing doses or environmental concentrations, over a set period. Below a concentration of 1 no mortalities are observed, but between concentrations of 1 to 2.5 the most susceptible individuals in the population succumb. Most of the population is affected by concentrations between 2.5 and 5, but some individuals are very resistant and it needs concentrations above 5 to cause mortality. So in any uniform population there will be a Gaussian distribution in terms of susceptibility to a toxic agent. It is expected that each agent will have a dose response curve characteristic of that substance.

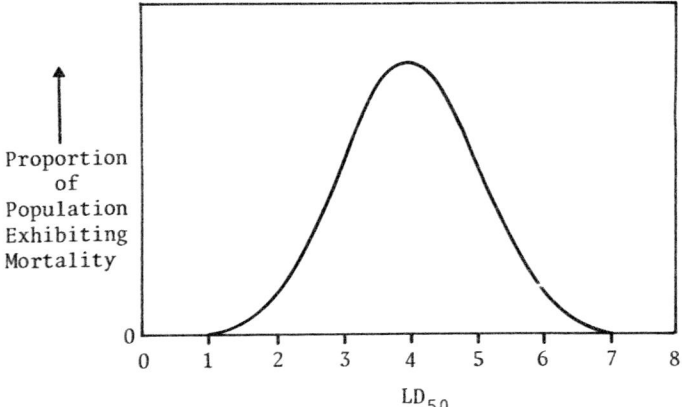

Fig. 2.3.1 Hypothetical lethal response curve of an animal population exposed to a toxicant.

Further examination of Fig. 2.3.1 shows that we can obtain a numerical measure of toxicity from the data. The concentration or dose which causes mortality to 50% of the organisms, LC_{50} and LD_{50} respectively, represents toxicity to the "average" organism and an estimate can be made of standard deviation. Often the Lethal Concentration and Lethal Dose are jointly referred to as the Tolerance Limit (TL) as this eliminates the need to consider the difference between a dose and exposure to a concentration. Toxicity to 84% (LC_{84}) or toxicity to 16% (LC_{16}) of the organisms can be also calculated if necessary.

It is important to note that within the same species a variety of factors may influence the measured toxicity. These include such characteristics as age, sex, size, general physical condition and environmental conditions.

Table 2.3.1 indicates the range of toxicities exhibited by a variety of substances towards mammals. This can be used to classify the toxicity of a substance according to a scale such as that illustrated in Table 2.3.2.

TABLE 2.3.1 Approximate LD50' of a Number of Chemical Agents for a Variety of Mammals

Agent	LD_{50} (mg/kg body weight)
ethyl alcohol	10,000
sodium chloride	4,000
morphine sulphate	900
DDT	100
nicotine	1
tetrodotoxin	0.10
"dioxin" (TCDD)*	0.001

*(From: Loomis, T.A. (1974), *Essentials of Toxicity*, 2nd Edn., Lea & Febiger, Philadelphia, p.18)

TABLE 2.3.2 General Classification of the Toxicity of Substances

Classification	LC_{50} or LD_{50} (mg/kg body weight or concentration)
extremely toxic	<1
highly toxic	1 - 50
moderately toxic	50 - 500
slightly toxic	500 - 5,000
practically non-toxic	5,000 - 15,000
relatively harmless	>15,000

(From: Loomis, T.A. (1974), *Essentials of Toxicity*, 2nd Edn., Lea & Febiger, Philadelphia, p. 19)

---

*2,3,7,8-tetrachlorodibenzo-p-dioxin

Aquatic toxicology tests, or bioassays, are conducted to evaluate the
toxicity of specific materials or effluents, to determine permissable
effluent discharge rates, to establish the relative sensitivity of various
aquatic organisms, and to identify effects of physical and chemical variables
such as temperature and pH on toxicity.

Two sorts of bioassay may be identified. One is the chronic, sub-lethal
test, which examines responses in essential life processes such as growth,
reproduction and changes in blood composition. The other is the acute-
lethality test, where the measured response is death. Chronic tests are
sometimes long-term while acute tests may take only a few days. With all
tests the test organisms should be as uniform in all characteristics as
possible and the environmental conditions should be constant during the
test period.

Simple static bioassays can be carried out in aquaria containing the test
organisms with different concentrations of the toxicants. This procedure
suffers from the disadvantages of a gradual loss of toxicant through
absorption by the organisms, and other mechanisms, as well as additional
toxic effects due to the wastes produced by the test organisms. More
accurate and consistent results are obtained using "flow-through" aquaria.
In this method the liquid in the aquaria is continually replaced by a
controlled inflow of fresh water and toxicant in the correct concentration.
Flow-through bioassays require complex equipment and also chemical monitoring
to check the toxicant concentration, therefore the following experiment
uses the static procedure.

The choice of test organism is of critical importance. Organisms selected
should be important from an ecological, commercial and recreational viewpoint
in the area where the test results apply. For practical reasons, they
should be easily obtainable in large numbers. But it is important to
remember that the conditions of the laboratory bioassay may differ from
that in the natural environment. For example the organism may have a
restricted area for movement in which factors such as dissolved oxygen,
temperature, salinity, pH and so on may differ from the test conditions.
In addition the size, age, sex and physical condition of the test organisms
may not be representative of the natural population. Thus there is a need
for caution in applying the results of bioassays to natural populations.

The chosen toxicant for this study is copper, as copper sulphate. Although
traces of copper are essential to living organisms higher concentrations
are toxic. Levels of 0.5 ppm in water are lethal to may algae, whilst most
fish succumb to a few parts per million. Major sources of copper pollution
are mining operations, industries such as electroplating works, and the use
of copper compounds as algicides and fungicides.

OBJECTIVES

The main objectives are:

1. to recognize the principles of a toxicological bioassay

2. to carry out a simple, small-scale static bioassay

3. to calculate 24- and 96-hr LC_{50} figures from mortality data.

MATERIALS

volumetric flasks, 100 and 500 ml
pipettes, 1 and 10 ml
5 litre pickling jars (5)
aquarium pump and tubing
analytical balance

copper sulphate
shrimp or other suitable aquatic organism* (about 80)

PROCEDURE

The bioassay covers about 5 days. It requires monitoring on frequent occasions and so adequate plans should be made ahead. Also, the test species must be collected and acclimatized in the laboratory, usually for a period of about one week. Native shrimp may be collected by dipnets and traps in local streams. Alternatively, brine shrimp eggs *(Artemia salina)* may be bought in an aquarium store and the shrimp cultivated in the laboratory.

First, a preliminary assay is conducted to determine the order of magnitude of copper concentration that is toxic to the shrimp.

1. Ensure that there is an adequate supply of dilution water. Tap water must be thoroughly dechlorinated by air scrubbing. If the tap water is hard, it should be diluted with an equal part of deionized water. Ensure also that the shrimp have been acclimatized to the dilution water. The shrimp should not be fed for one day before the test.

2. Clean and rinse thoroughly five 5 litre jars. Fill with dilution water and add copper sulphate solution to reach the following concentrations: 0, 0.01, 0.1, 1 and 10 mg/l as copper. The zero copper water will serve as a control.

3. Place two shrimp in each jar, selecting shrimp that are of similar size and the same sex. Observe the shrimp for a few minutes, and replace those that show obvious signs of damage through handling.

4. After 24 hours have elapsed, note the mortalities. There should be no deaths in the control. Ideally, there should be a copper concentration at and above which there is 100% mortality and below which there is zero mortality.

A detailed assay should be conducted to span a range of copper concentrations from that causing zero mortality to that causing complete mortality. Concentrations following a logarithmic series are most appropriate and may be taken from the following list:

   1.0, 1.8, 3.2, 5.6, 10, 18, 32, etc.

Thus a tenfold range may require five concentrations.

---

* A satisfactory organism should be suitable for collection or breeding in large numbers; of the smallest size which allows observation and handling and should exhibit a clear death point. Government regulations and conservation needs should be obeyed.

5. Fill several 5 litre jars with dilution water. Adjust the copper concentrations to the required values, leaving one jar to serve as a control.

6. Stock each tank with ten shrimp of the same sex and roughly the same size as used in the preliminary assay taking care not to injure them in the process.

7. Run the assay for 96 hours (four days).. Do not feed the shrimp over this time. Record the shrimp mortalities at known time intervals. A convenient set of intervals is:

```
15, 30, 70 minutes,
2, 4, 8 and 14 hours,
1, 2, 3 and 4 days.
```

Other, more convenient sets of intervals may be chosen.

8. Remove the shrimp as soon as they die, even if this is between the chosen observation times. Shrimp should be considered dead if there is no respiratory or other movement, or no response to gentle prodding.

ANALYSIS OF DATA

Several methods are used to analyse bioassay data. One method is illustrated below with a set of mortality data for groups of 10 shrimp exposed to five copper concentrations for 4 days (see Table 2.3.3). The percentage mortality in each test group is plotted against the copper concentration for each observation time. Fig. 2.3.2 illustrates the plot for the 14-hr observation.

TABLE 2.3.3 Shrimp Mortality Data

Copper concentration mg/l	Number of tests animals surviving after:								
	30min,	1,	2,	4,	14,	24hr,	2,	3,	4days
10	9	7	4	2	0	0	0	0	0
5.6	10	9	7	5	2	1	1	0	0
3.2	10	10	9	7	5	4	3	2	2
1.8	10	10	10	9	8	7	6	5	5
1.0	10	10	10	10	10	10	9	9	9

Fig. 2.3.2 shows that the data may be fitted to a sigmoid curve. This type of curve is expected of natural populations. Most shrimp behave close to the "average", and succumb to a narrow range of toxicant concentration. However, a few shrimp display particular sensitivity or resistance to the toxicant.

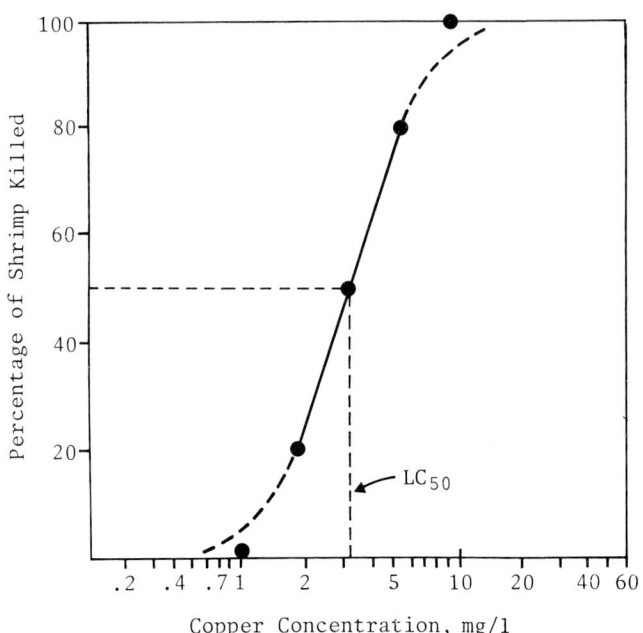

Fig. 2.3.2 Plot of mortality *versus* toxicant concentration at 14 hours. Note the logarithmic scale for the concentration. The dashed curve would fit ideal figures. The solid line is better fitted to data for a limited number of animals which behave similarly to the average organism. From the solid line is obtained a 14-hr LC_{50} of 3.2 mg/l of Cu.

In practice, it is easier to fit a straight line to experimental data. Fig. 2.3.2 shows that a straight line can be fitted to the points closest to the 50% median, ignoring the points for 0 and 100% mortality. From this line can be read off the 14-hr LC_{50} of 3.2 mg/l of Cu.

Other LC_{50} values may be estimated for each observation time. These values are plotted against time as the experiment progresses, to obtain a "toxicity curve" as shown in Fig. 2.3.3. From this curve can be read off the 24- and 96-hr LC_{50} which in this case are 2.5 mg/l and 1.8 mg/l of copper, respectively.

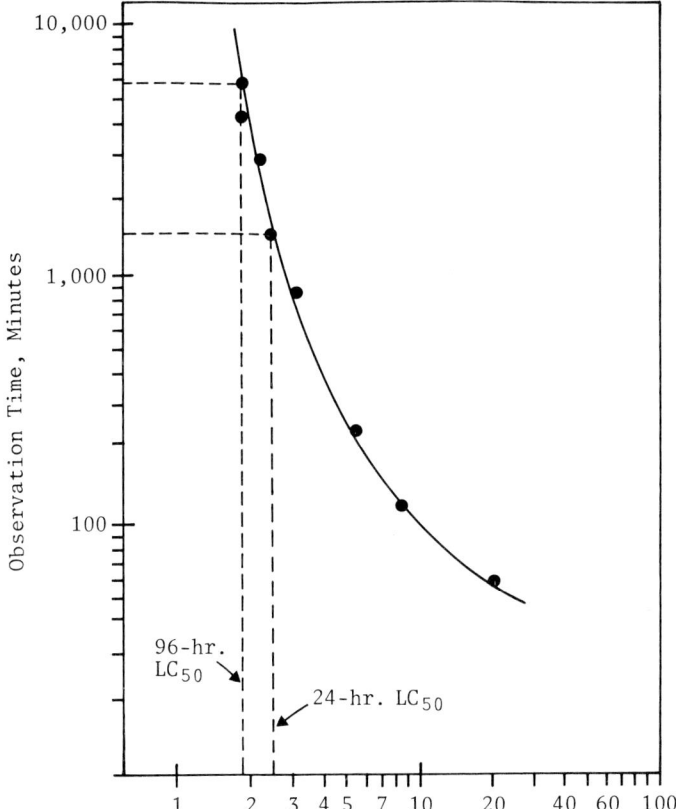

Fig. 2.3.3 Plot of LC$_{50}$ *versus* observation time. This plot should be drawn as the bioassay proceeds. In this case, deaths are still occurring at 96 hours. Longer experiments may indicate a concentration at which acute lethality ceases, a "threshold lethal concentration".

QUESTIONS

1. What is the LC$_{50}$ at 24 hours and 96 hours for the test organisms under the conditions used?

2. What are the government regulations for copper concentrations in aquatic areas? How do these compare with the LC$_{50}$ data obtained from your experiment?

3. What is meant by "threshold lethal concentration"? Can you estimate a threshold lethal concentration for copper from your bioassay data?

4. What is meant by *synergysm* and *antagonism* in terms of pollutant bioassay? How does water hardness influence the toxicity of dissolved copper.

BIBLIOGRAPHY

American Public Health Association, American Water Works Association and Water Pollution Control Federation (1975), *Standard Methods for the Examination of Water and Wastewater*, 14th Edition, APHA; Washington.

Glass, G.E. (Ed.) (1973), *Bioassay Techniques and Environmental Chemistry*, Ann Arbor Science, Ann Arbor.

Katz, M. (1971), Toxicity Bioassay Techniques Using Aquatic Organisms. In: L.L. Giacco (Ed.), *Water and Water Pollution Handbook*, Volume 2, Marcel Dekker, New York.

Sprague, J.B. (1978), The ABC's of Pollutant Bioassay Using Fish. In: J. Cairns and K.L. Dickson (Eds.), *Biological Methods for the Assessment of Water Quality*, Special Technical Publication No. 528, American Society for Testing and Materials, Philadelphia.

2.4 LEAD IN HOUSEHOLD PAINT

INTRODUCTION

Lead has an extensive history in human experience. Lead has been mined and worked for thousands of years. Its widespread early use is accounted for by its ease of refinement from natural ores and its properties of ductility and high resistance of corrosion. The history of health hazards associated with its use in equally extensive. Lead poisoning, or plumbism, was first described by the Greek poet-physician Nicander more than 2000 years ago.

The toxicology of lead is well understood today. It is absorbed by two major routes, the alimentary and the respiratory. Under normal conditions, more then 90% of lead retained in the body is in the skeleton. The remainder is excreted back into the gut in the bile and is also excreted in urine, sweat, hair and nails. Lead is highly cumulative and only a small amount needs be mobilized to add appreciably to the amount in the soft tissue pool. Lead stores in bone can persist for months and even years so that acute lead poisoning can occur long after exposure to abnormal amounts or lead has ceased.

Lead poisoning results from the high levels of lead in the soft tissues. Acute lead poisoning is possible if the concentration in blood rises above about 0.8 ppm. It most often strikes the blood, kidney and nervous system, and can result in anaemia, chronic nephritis and convulsions. Lower levels of lead poisoning can lead to such chronic effects as malaise, muscle aches, headache plus mild anaemia and shortened erythrocyte life spans.

Modern sources of lead contamination are widespread. They range from lead mining and smelting operations, battery manufacture, to automobile emissions (see Table 2.4.1). A particularly insidious source of contamination is the use of lead pigments in paint. These pigments include lead chromate (red, orange), lead oxide (red) and lead sulphate (white).

TABLE 2.4.1 Average Content of Lead in U.S. Gasoline

Gasoline	Concentration (g/gallon)	
	1970	1971
regular	2.43	2.22
premium	2.81	2.67
low lead & no lead	0.75	0.75

(From: Waldbott, G.L. (1978), *Health Effects of Environmental Pollutants*, C.V. Mosby, St. Louis, p. 143)

In most countries government regulations severely limit the use of lead pigments in paint. This stems from the connection found between the high incidence of lead poisoning in children and the widespread use of lead-based paint on wooden homes, Children with the habit of eating non-food items (pica) were picking up flakes of lead paint. Many older buildings are still coated with lead containing paints. Some children have suffered acute poisoning from drinking the rainwater caught on building surfaces.

The estimation of lead in environmental samples is commonly performed using atomic absorption spectroscopy (A.A.S.). This technique exploits the ability of atoms to emit and absorb light at certain discrete wavelengths. Fig. 2.4.1 is a schematic diagram of an atomic absorption spectrometer. It includes a hollow cathode or vapour discharge lamp containing the metal of interest. The sample being investigated, usually in solution in water, is drawn into a high temperature flame where it is atomized. The light from the lamp passes through the flame where it is partially absorbed by the excited atoms. The attenuated radiation then passes a slit into a monochromator which selects the particular radiation band. Finally, the light intensity is measured by a conventional photomultiplier tube and amplifier.

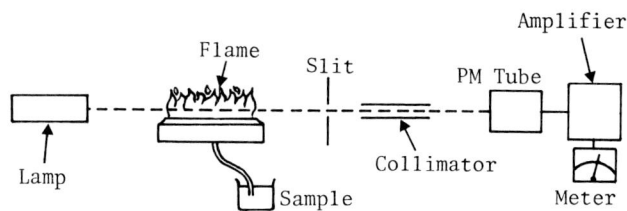

Fig. 2.4.1 Schematic diagram of an atomic absorption spectrophotometer.

OBJECTIVES

The main objectives of this activity are:

1. to identify basic principles of atomic absorption spectroscopy

2. to work up environmental samples for A.A.S. analysis of lead

3. to determine the lead content of the sample using a standard curve.

MATERIALS

Atomic absorption spectrophotometer
muffle furnace
hot plate
volumetric pipettes, 1,2,5 & 10 ml
graduated pipette, 10 ml
volumetric flasks, 100 ml
centrifuge

Concentrated nitric acid
stock solution of lead,
    1000 µg/ml
samples of paint

PROCEDURE

The following method is applied to housepaint scrapings. However, other pigmented specimens may be used, such as pencil enamel or coloured confectionary wrappings. Several specimens should be examined so as to obtain a range of samples.

1. Weigh duplicate samples of about 0.2 g of paint scrapings into two small silica crucibles. Use a third crucible for a blank determination. Weigh whole wrappers if these are chosen.

2. Carefully ignite and ash in a muffle furnace set to reach a miximum temperature of 600ºC over 1 hour.

3. Remove each crucible and allow to cool. Add 2 ml of concentrated nitric acid, then warm gently on a hotplate.

4. Transfer the mixture into a 100 ml volumetric flask, rinsing the crucible thoroughly with mineralized water. Make up to the mark with mineralized water, shake and let stand. If the solution remains excessively cloudy, centrifuge a portion for 5 minutes.

5. Aspirate the solution into the atomic absorption spectrophotometer set at 217.0 nm and using continum background correction. Record the absorbance. If the meter is off-scale, dilute the sample with 2% nitric acid till an absorbance is obtained. With no interferences, the absorbance for the blank should be negligable.

6. Dilute the stock solution of lead to obtain several standard solutions of 5 - 25 ppm Pb in 2% nitric acid. Measure their absorbance, then plot a standard curve of absorbance versus lead concentration to obtain a straight line.

7. Read off the lead concentration of the sample solutions. Calculate the lead content of the original paint samples and express as microgram per gram.

QUESTIONS

1. What is the range of lead concentrations found in the samples?

2. What is the "acceptable" lead content for the samples analysed? (Use as a guide the Health Acts, the Food and Drug Regulations, or any other state, national or international standards). Does your investigation indicate compliance with these levels?

3. Certain sequestering agents are administered to counter lead poisoning. What are some common agents, and how do they act?

BIBLIOGRAPHY

Bertagnolli, J.F. and Katz, S.A. Coloured gift Wrappings as a source of Toxic Metals, *Intern. J. Environ. Anal. Chem.*, *6*, 321 (1979).

Anon (1973), *Environmental Health Aspects of Lead*, Commission of the European Communities, Luxembourg,.

Waldbott, G.L. (1978), *Health Effects of Environmental Pollutants*, 2nd Edn. Mosby, Saint Louis.

Anon (1977), *Environmental Health Criteria, No. 3 Lead*, World Health Organisation, Geneva.

2.5 ATMOSPHERIC POLLUTANTS

INTRODUCTION

The air we breathe can be described as a colourless, odourless mixture of gases composed of 78% Nitrogen, 21% Oxygen, 0.3% Carbon Dioxide and about 0.7% rare gases such as Krypton, Xenon and Argon. In addition there are highly variable amounts of water present either as vapour or as droplets and many other chemical substances in trace amounts. Substances present in the air are regarded as pollutants if they are present in concentrations toxic to man, animals or plants, have an odour or in some other way irritate our senses; obscure visibility or damage property.

Table 2.5.1 lists some pollutant gases with a comparison between their concentration in a natural environment and a polluted zone. As well as these gases, which are largely the products of combustion, there are other substances such as heavy metals, halogens or organic halogen compounds, and particulate matter. The natural background concentrations are averages over longer times, whereas the maxima in polluted areas are peak values during episodes of extreme pollution. The polluted areas show concentrations of most trace substances approximately two orders of magnitude higher than the natural background. Table 2.5.2 summarises some air quality criteria recently set by the U.S.A. and W.H.O. for several of these pollutants.

TABLE 2.5.1 Trace Substances Which Occur in Air

Substance	Natural background concentration, $\mu g/m^3$	Concentration in polluted air, $\mu g/m^3$
Methane	1000	1500 - 2500
Non-Methane Hydrocarbons	3	50 - 500
Carbon Monoxide	100	2500 - 10^5
Sulphur Dioxide	5	100 - 2000
Nitrogen Dioxide	2	100 - 1500
Ozone	30	100 - 1500

The air pollutants examined in this activity are nitrogen oxides, carbon monoxide and oxidants. The first two are produced in combustion processes, while oxidants are formed by photochemical reactions that take place in the atmosphere.

Nitrogen oxides are often termed NO_x to represent the combined contributions of nitrogen monoxide and nitrogen dioxide, NO and NO_2. Although vast quantities are produced by natural biological reactions, the resultant concentrations are low due to the wide dispersal. High local concentrations are formed from nitrogen and oxygen in the atmosphere at elevated temperatures such as those that accompany the burning of fossil fuels. Automobile exhausts contribute significantly to atmospheric levels of NO_x, the primary emission being nitrogen monoxide:

$$N_2 + O_2 \rightarrow 2NO$$

Once formed, nitrogen monoxide is slowly oxidized to the dioxide, the period of oxidation being 4-6 days:

$$2NO + O_2 \rightarrow 2NO_2$$

Nitrogen dioxide has a sharp, irritating odour, is toxic and can dissolve in water to form a solution of nitric acid:

$$3NO_3 + H_2O \rightarrow 2HNO_3 + NO$$

A most important feature of nitrogen dioxide is its role as a precursor in photochemical smog. Under the action of strong sunlight, nitrogen dioxide dissociates forming highly reactive atomic oxygen:

$$NO_2 + \text{sunlight} \rightarrow NO + O$$

This reactive species is involved in a complex series of reactions involving oxygen, hydrocarbons, nitrogen oxides as well as other compounds:

$$O + O_2 \rightarrow O_3$$

$$O_3 + \underset{\substack{\text{partly} \\ \text{R burnt} \\ \text{hydrocarbon}}}{\overset{CH_3}{\underset{CH}{\overset{|}{\underset{\|}{CH}}}}} \rightarrow \rightarrow \underset{}{\overset{CH_3}{\underset{O}{\overset{|}{\underset{\|}{CH}}}}} + \text{other oxygen containing products}$$

$$\underset{}{\overset{CH_3}{\underset{O}{\overset{|}{\underset{\|}{CH}}}}} \xrightarrow[\text{free radical reactions}]{\text{photochemical}} \rightarrow \rightarrow \rightarrow \rightarrow \rightarrow \underset{}{\overset{CH_3}{\underset{O}{\overset{|}{\underset{\|}{C-O-O^\cdot}}}}}$$

$$\underset{}{\overset{CH_3}{\underset{O}{\overset{|}{\underset{\|}{C-O-O^\cdot}}}}} + NO_2 \rightarrow \underset{\substack{\text{peroxyacetylnitrate} \\ \text{(PAN)}}}{\overset{CH_3}{\underset{O}{\overset{|}{\underset{\|}{C-O-O-NO_2}}}}}$$

Although very simplified, these equations illustrate the formation of several pollutants such aldehydes, ozone and PAN. Aldehydes can condense to form aerosols, which limit visibility and are toxic. Ozone and PAN are members of a group of oxidants which are extremely toxic to plants, cause oxidative damage to many materials such as fabrics, plastics and rubber, and are very powerful lachrymators or eye irritants.

On a mass basis, carbon monoxide is by far the most important atmospheric pollutant in zones of high pollution. Sources of carbon monoxide pollution are almost entirely the incomplete combustion of carbonaceous material.

TABLE 2.5.2 Some National and International Air Quality Criteria*

	NO_2	Oxidants	SO_2	CO	Particulates
U.S.A.	100 (year)	160 (1hr)	80 (year)	10^4 (8hr) 4×10^4 (1hr)	75 (year) 250 (24hr)
W.H.O.	190-320 (1hr)	60 (8hr) 120 (1hr)	60 (year)	10^4 (8hr) 4×10^4 (1hr)	40 (year)

*$\mu g/m^3$ - figures in parentheses are exposure times.

Worldwide, there are much larger natural sources. Methane produced by the decomposition living matter can react with hydroxyl radicals to form methyl radicals and water:

$$CH_4 + .OH \rightarrow .CH_3 + H_2O$$
$$\downarrow \text{(several steps)}$$
$$CO$$

This series of reactions produces an estimated 80% of the carbon monoxide in non-urban atmospheres.

The principal detrimental effect of carbon monoxide on humans and animals is the interference with oxygen transfer through the body. By forming a very stable complex with haemoglobin (Hb) in the red blood cells, carbon monoxide displaces oxygen and prevents the latter from being carried through the bloodstream:

$$HbO_2 + CO \rightleftarrows COHb + O_2 \quad (K \sim 210)$$

The normal level for carboxyhaemoglobin (COHb) in the blood is about 0.5%. For concentrations of carbon monoxide below 100 ppm in the inhaled/exhaled air the equilibrium level of COHb follows roughly the following equation:

$$\% \text{ COHb in blood} = 0.16 \times [ppm\ CO] + 0.5 \quad \text{(Equation 2.5.1)}$$

The percentage of heamoglobin immobilised by a wider range of atmospheric concentrations of CO is shown in Fig. 2.5.1.

The analysis of nitrogen oxides and oxone in air may be conveniently performed using "wet" colourmetric methods. Both involve drawing the air sample through colour-forming reagent solutions. For nitrogen dioxide, hydrolysis to nitrate and the subsequent reaction with sulphanilic acid and N-(1-naphthyl)ethylenediamine in acid medium yeilds a pink coloured azo-dye complex. Ozone reacts with iodide ion in neutral buffer solution according to the reaction:

$$O_3 + 3I^- + H_2O \rightarrow I_3^- + 2OH^- + O_2 \quad \text{(Equation 2.5.2)}$$

The triiodide ion liberated has an intense yellow colour.

Other gases may interfere with these colourimetric determinations. With ozone in particular, reducing gases such as sulphur dioxide and hydrogen sulphide will negatively interfere. Triiodide ion will also be liberated by other oxidants such as PAN, peroxy compounds, organic nitrites, and halogens. Because of the non-specific nature of the iodide method, the result obtained is usually given as Total Oxidants, with concentration units in terms of ozone which is normally the most abundant atmospheric oxidant.

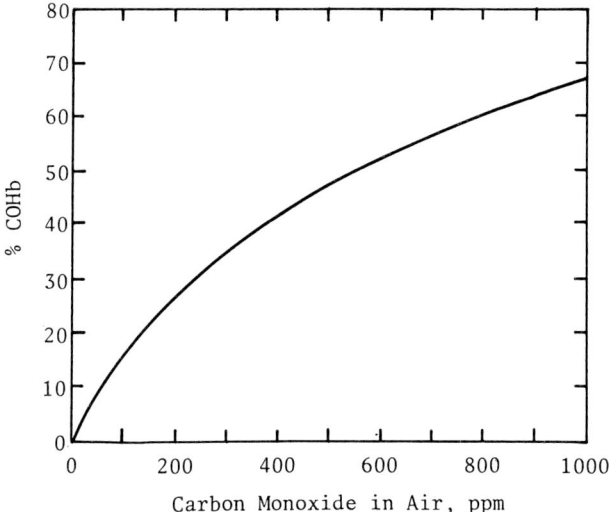

Fig. 2.5.1 Equilibrium Percentage of Haemoglobin immobilized by carbon monoxide over a range of ambient CO concentrations.

OBJECTIVES

The main objectives are:

1. to determine the ambient levels of nitrogen dioxide and photochemical oxidants by wet chemical methods

2. to determine the carbon monoxide of content human breath.

MATERIALS

Interscan CO analyzer or equivalent
calibration gas, c. 50 ppm CO
air sampling trains
plastic bags
spectrophotometer
pipettes
volumetric flasks

Buffered KI oxidant absorbing
  reagent
(13g KH_2PO_4, 14.2 g Na_2HPO_4,
  10 g KI in one litre of
  distilled water)
standard 0.05M iodine solution
potassium permanganate phosphoric
  acid sulphuric acid
  solution

standard (1.08 g/l) $NaNO_2$
solution
NO_2 absorbing reagent

PROCEDURE

Total Oxidants

The total oxidant concentration of ambient air is determined by reaction with buffered potassium iodide solution and the measurement of released iodine at 352 nm. If reducing gases such as SO_2 interfere significantly with the assay, the air sample can be pretreated according to the method outlined by Perry and Young (see Bibliography).

1. Set up an air sampling train so as to contain one midget sampling impinger.

2. Place 10 ml of oxidant absorbing reagent into the midget impinger. To prepare the absorbing reagent, dissolve successively 13.6 g of KH_2PO_4, 14.2 g of Na_2HPO_4 and 10g of KI in one litre of demineralized water. The reagent should be kept for at least one day before use and is stable for several weeks if stored refrigerated in a dark glass bottle.

3. Draw air at 1-2 litres per minute through the sample impinger. Stop after about 30 minutes, and record the volume of air sampled.

4. Let the exposed absorbant solution stand for a further 30 min, then place some of the exposed reagent in a 1 cm cuvette and measure the optical absorbance at 352 nm. Use unexposed reagent as the reference.

5. Prepare some calibration iodine solution of the oxidant measurement. Take 2 ml of standard 0.05 M iodine solution and dilute to 100 ml with demineralized water. Take 5 ml of this solution and further dilute to 100 ml with oxidant absorbing reagent. This provides a calibration iodine solution equivalent to 1.92 µg of ozone per ml according to the stoichiometry of Equation 2.5.2.

6. To obtain a range of values, add graduated volumes of the calibration solution to a series of 10 ml volumetric flasks and dilute to the mark with absorbing reagent. Aliquots of 2, 4 and 6 ml are convenient. Read the optical absorbances at 352 nm and plot them against ozone concentration in µg per 10 ml of absorbing reagent. The plot should be linear passing through the origin.

7. Read off the total O_3 content of your sample from the calibration curve. The oxidant concentration in the air sample is given by

$$\text{Oxidant } (\mu g/m^3) = \frac{\text{total µg } O_3 \text{ per 10 ml of absorber}}{\text{volume of air sampled in cubic metres}}$$

Nitrogen Oxides

The colourimetric method is specific for nitrogen dioxide only. It may be extended to include the contribution from nitric oxide if the air is first drawn through an oxidising solution that converts NO to NO_2.

1. Arrange an air sampling train so as to contain one or two midget sampling impingers. A convenient arrangement is shown in Fig. 2.5.2.

2. If NO is to be measured as well as NO_2, fill the first impinger with 20 ml of freshly prepared acidic potassium permanganate solution (about 0.5 g of $KMnO_4$ in 20 ml of a mixture of 10 ml of 18 M H_2SO_4 and 100 ml of 60% H_3PO_4).

3. Place 25 ml of NO_2 absorbant reagent in the second impinger. The absorbent reagent is made by dissolving 10 g of sulphanilic acid, 0.1 g of N-(1-naphthyl)-ethylmediamine dichlorohydrate, 20 ml of propan-1-ol, and making up to 2 litres with demineralized water. The absorbant reagent should be stored in dark bottle and should be free of any pink colour.

Fig. 2.5.2 Commercial gas sampler comprising battery driven vacuum pump, rotameter, air velocity controls and two glass midget impingers. The oxidant solution (dark) is prevented from contaminating the aborder reagent by a glass-fibre filter. (Photo: John Watson).

4. Sample air at 1-2 litres per minute until a noticeable pink colour has developed.

5. After sampling, let stand for 15 minutes for colour development. Measure the optical absorbance at 550 nm against an absorber reagent reference. If the absorbance is too high, dilute using reagent solution.

6. Prepare a calibration solution using sodium nitrite in place of NO_2. It has been found empirically that 0.72 mole of $NaNO_2$ gives the same colour as 1 mole of NO_2, so that 1.08 g of $NaNO_2$ is equivalent to 1 g of NO_2.

7. Obtain the standard solution containing 1.08 g of dry $NaNO_2$ per litre. Dilute 10 ml to one litre with demineralized water to obtain a solution where 1 ml contains the equivalent of 10 µg of NO_2. Transfer 0.5, 1, 1.5 and 2 ml volumes of this dilute solution to a series of 25 ml volumetric flasks and make up to the marks with absorbant reagent.

8. Let the colour develop for 15 minutes, then read the absorbances at 550 nm. Plot the absorbance *versus* µg of NO_2 per 25 ml of absorber reagent, to obtain a straight line passing through the origin.

9. Read off the NO_2 content of your sample solution from the calibration curve. The NO_2 or NO_x concentration in air is then given by:

$$NO_2 \; (\mu g/m^3) = \frac{\mu g \text{ of } NO_2 \text{ in 25 ml of absorber}}{\text{volume of air sampled in cubic metres}}$$

Carbon monoxide in human breath

Carbon monoxide is conveniently measured using a commercial electrochemical monitor, such as the Interscan COtector or its equivalent.

1. Allow the CO monitor sufficient warm-up, then set the internal electrical zero. Calibrate the instrument against a suitable standard, about 50 ppm in synthetic air.

2. Collect breathe samples in plastic bags, then connect to the inlet tube. Record the values obtained.

3. Estimate the carboxyhaemoglobin content of each subject's blood using Equation 2.5.1.

QUESTIONS

1. What are the concentrations of oxident and nitrogen oxides found?

2. What range of carboxyhaemoglobin concentrations were found in the blood of subjects examined? If there are large variations what is the cause?

3. How do your data for nitrogen and oxides and oxidants conform to local standards or those set down by the World Health Organization or the U.S. Environmental Protection Agency?

4. What are the standards for CO in air? What are the carboxyhaemoglobin levels at which the first signs of CO poisoning appear?

BIBLIOGRAPHY

Anon, W.H.O. Environmental Health Criteria for Oxides of Nitrogen, *Ambio*, *6*, 290 (1977).

Anon (1976), Manual on Urban Air Quality Management, *W.H.O. Regional Publications European Series No. 1*, World Health Organization, Copenhagen.

Newill, V.A. (1977), Air Quality Standards. In: A.C. Stern (Ed.) *Air Pollution*, Volume 5, 3rd Ed., Academic Press, New York.

Perry, R. and Young, R.J. (Eds.) (1977), *Handbook of Air Pollution Analysis*, Chapman & Hall, London.

PART 3

Food Additives and Contaminants

3.1 AFLATOXINS IN GROUNDNUTS

INTRODUCTION

Aflatoxins are a group of toxic fluorescent compounds produced by fungal growth on foods. Fungi of the genus *Aspergillus* are the main toxin producers, including the common mould, *Aspergillus flavus*.

Several aflatoxins have been identified, including a group of difuranocoumarin derivatives which are denoted as B_1, B_2, G_1 and G_2:

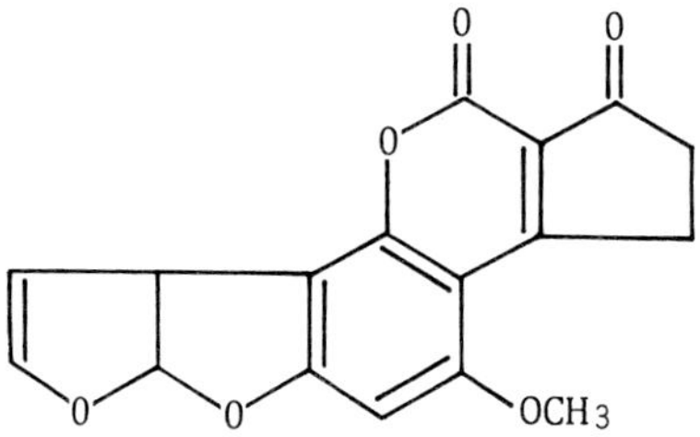

Aflatoxin B_1

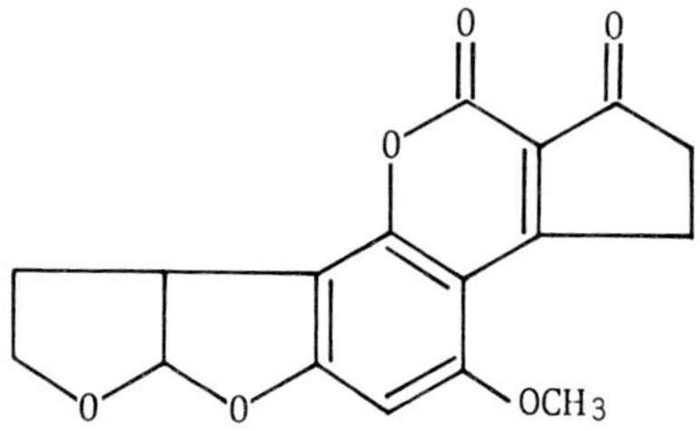

Aflatoxin B_2

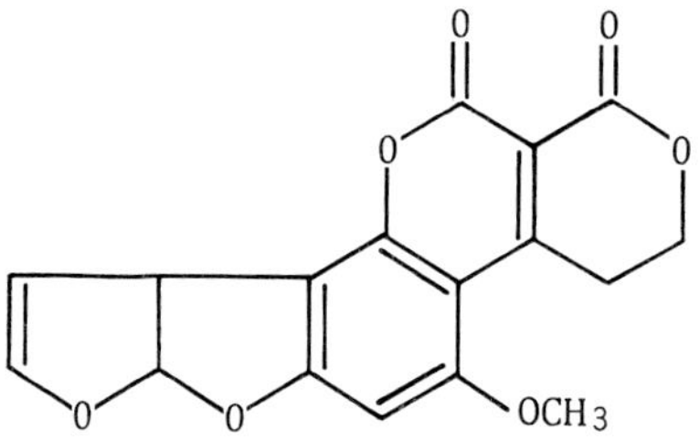

Aflatoxin G_1

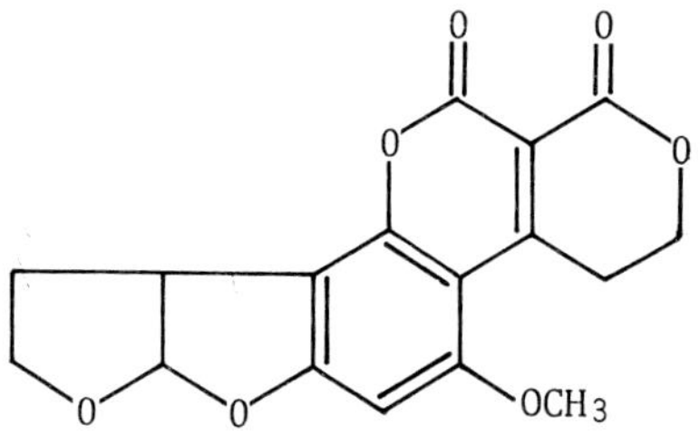

Aflatoxin G_2

The notation B and G refers to the blue or green fluorescence shown by the toxins. Aflatoxins B_2 and G_2 are dihydro derivatives of B_1 and B_1, respectively. Of this group, Aflatoxin B_1 is the most common in naturally contaminated foods.

The current interest in the aflatoxins stems largely from their association with diseases in animals and man. For example, in 1960, about 100,000 turkeys in Britain died of acute poisoning, apparently due to their being fed on grain contaminated with *A. flavus*. Heavy aflatoxin contamination has been found in food samples taken from parts of Africa and South-East Asia, areas where there is a high incidence of liver cancer among the population. Experimental studies with rats have show aflatoxins to be extremely potent liver carcinogens, Aflatoxin B_1 being the most potent of all known substances.

TABLE 3.1.1 Aflatoxin Limits in Different Countries 1976

Country	Commodity	Aflatoxin limit (µg/kg)
Canada	nuts and their derived products	15[b]
France	animal feed	700
India	food	30
	groundnut flour for food use	120
	groundnut cake (export)	60-120
Japan	all foods	10
	groundnuts cake for animal feed mixes	1000
Netherlands	foods and feeds	5
Poland	all foods and feeds	5
Rhodesia	groundnuts	25
	animal feed	50-400
Sweden	all foods, particularly Brazil nuts, groundnuts, groundnut butter, raw materials for further processing in Sweden	5
		20
U.K.	confectionery groundnuts	50
	groundnut flour for animal feeds	0-500[c]
USA	confectionery groundnuts	20
	all foods and animal feeds	20-25[b]
	produce for processing into mixed feed	50
EEC countries	complete feed for cattle, sheep and goats (with the exception of dairy animals, calves and lambs)	50
	complete feed for pigs and poultry (with the exception of infant pigs, chicks, ducklings and turkeys)	20
	animal feed supplements for dairy animals	20
	other complete feeds	10

[a] Aflatoxin B_1,

[b] Total Aflatoxins B_1, B_2, G_1 and G_2

[c] EEC limits may apply

Adapted from: Anon (1977), *Conference on Mycotoxins*, FAO/WHO/UNEP, Nairobi, Kenya.

Aflatoxin contamination is a problem with commodities such as nuts and grains, particularly if the conditions during harvest or storage are warm and moist. Peanuts and peanut products have been the most widely studied, but aflatoxins have been detected in a wide variety of foods. These foods include brazil nuts, pistachio nuts, almonds, walnuts, pecans, filberts, cottonseed, copra corn, grain, sorghum, rice and figs.

Many countries have established regulatory limits for aflatoxins in susecptible foods, and most such limits have been set close to or at the limit of analytical detection, reflecting the view that there is no known safe level of human exposure. Table 3.1.1 lists some limits for human food and animal feed.

OBJECTIVES

The main objectives are:

1. to assay for the presence of aflatoxin in groundnuts using solvent extraction and column chromatography

2. to identify the aflatoxin using thin layer chromatography.

MATERIALS

Blender
culture tubes, 18 x 150 mm
filter funnel
pastuer pipettes or disposble
  plastic minicolumns (available from
  Applied Science Laboratories, Inc.)
filter paper 9 cm
U.V. viewer, 360nm
TLC plates, non-fluorescent silica gel
glass vial and cap
graduated pipettes, 3 & 10 ml
microsyringes, 10 and 100 μl
chromatography tank

Raw groundnuts
sodium chloride
zinc acetate
acetic acid
methanol
hexane
chloroform
acetone
florisil & alumina (optional)
mixed aflatoxin standard
  (B_1 and G_1, 1 μg/ml)
  B_2 and G_2, 0.2 μg/ml)
trifluoroacetic acid

PROCEDURE

A rapid preliminary assay is performed by extraction/cleanup of the sample with solvent partition, small-scale column chromatography. If the presence of aflatoxins is indicated by their characteristic fluorescence, thin layer chromatography is then used to identify individual compounds.

CAUTION: *Aflatoxin concentrates are very hazardous and should be handled with protective gear in fume cupboards. Contaminated materials and spillage should be swabbed with 5% NaOCl solution.*

1. Blend 100 g of nuts with 200 ml of methanol-water (80:20), 100 ml of hexane and 4 g of sodium chloride at high speed for 1 minute. Centrifuge for 10 minutes to allow the mixture to separate.

2. Filter 10 ml of the methanol-water phase into a culture tube using fast filtering paper. Add 10 ml of mixed salt solution (60 g sodium chloride, 60 g zinc acetate and 1.5 ml glacial acetic acid in 400 ml of distilled water). Cap the tube and shake vigorously for 5-10 seconds.

3. Filter 15 ml of the mixture through a 9 cm glass-fibre filter into a second culture tube. Add 3 ml of chloroform, cap the tube, and shake the contents for 10 seconds. Allow the two layers to separate.

4. Prepare a small chromatography column using a Pasteur pipette. Plug the base of the pipette with a 5 cm wad of cotton wool. Add Florisil to make a column 15 mm long. Form a level surface, then add alumina to a depth of 15 mm. Firmly plug the top of the column with another wad of cotton wool. Alternatively, prepacked disposable plastic aflatoxin assay minicolumns are available commercially from Appied Science Laboratories, Inc.

5. Pipette 1 ml of the chloroform layer on to the top of the minicolumn and draw through by applying vacuum. After the chloroform has been drawn through, elute with 12 ml of methanol using the vacuum. Continue applying vacuum for another 2 minutes, till all the methanol has evaporated from the column.

6. View the minicolumn under long-wave U.V. light (360 nm). A blue fluorescent band at the Florisil-alumina interface indicates the presence of aflatoxins.

Samples proving positive in the preliminary assay can be treated *via* thin layer chromatography to identify the aflatoxin components present.

7. Place 1 ml of the chloroform solution formed as above in a small glass vial. Evaporate off the solvent with warming under a gentle stream of air. Cap the vial with a hollow polyethylene stopper.

8. Add 0.1 ml of chloroform to the vial *via* a microlitre syringe. Shake vigorously to dissolve. Pierce the polyethylene stopper to accommodate the needle of a 10 µl syringe.

9. Spot 10 µl of the solution at about 4 cm from the bottom edge of a 5x20 cm TLC plate. Spot alongside 5 µl of the mixed aflatoxin standard (B_1, B_2, G_1, G_2).

10. Develop the plate in a tank containing 50 ml of acetone-chloroform (1 + 9). When the solvent front has run 12-14 cm above the origin, remove the plate and allow the solvent to evaporate at room temperature.

11. View the plate under long-wave U.V. light. There should be a pattern of four fluorescing spots for the standard mixture. In order of descending R_f they are B_1, B_2, G_1 and G_2. Note the small colour difference, the bluish fluorescence of "B" contrasting with the slightly green of the "G" aflatoxins. Identify aflatoxins in the food extract by comparision.

Some food samples contain compounds with fluorescent and chromatographic properties similar to that of B_1. In the analysis of such samples, a chemical confirmatory test should be carried out.

12. Divide a silica gel TLC plate into two halves by scoring a vertical line. On one half of the plate, spot 10 µl of sample solution and 5 µl of B_1 standard. Place one drop of trifluoroacetic acid on each of these two spots and let it react for 5 minutes. Blow warm air on the bottom of the plate for about 10 minutes to dry.

13. Place sample and B_1 standard spots on the other half of the plate but do not add trifluoroacetic acid or heat. Develop the chromatogram (15 + 85).

14. Examine the developed plate under U.V. light. Unreacted aflatoxins should appear near the top on the side with no trifluoroacitic acid. A blue fluorescent derivative should appear at R_f a quarter that of B_1 with trifluoroacetic acid.

QUESTIONS

1. Which aflatoxins were detected in the samples examined?

2. Aflatoxins other than substituted coumarins are known. What is the structure of Patulin and from which mould is it formed?

BIBLIOGRAPHY

Rodricks, J.V. (Ed.) (1976), Mycotoxins and Other Fungal Related Food Problems, *Advances in Chemistry Series* 149, American Chemical Society, Washington, D.C.

Goldblatt, L.A. (Ed.) (1969), *Aflatoxin: Scientific Background, Control and Implications*, Academic Press, New York.

Holaday, C.E. and Lansden, J. Rapid Screening Method for Aflatoxins in a Number of Products, *J. Agric. Food Chem.*, 23, 1134 (1975).

3.2 FOOD ADDITIVES

INTRODUCTION

A variety of chemical substances are added to foods to improve their flavour, appearance, nutritive value and keeping qualities. These additives are classed according to their function and usage as preservatives, antioxidants, colours, flavours, sweeteners, emulsifiers, acidulants, aerators, humectants, free-running agents, vitamins and minerals. Two of these classes, preservatives and colours, are under investigation in this activity and are discussed below.

1. Preservatives

Benzoic acid and its sodium salts are among the most widely used antibacterial agents used in foods. Many berries (e.g. raspberries) contain appreciable amounts (up to 0.05%) of benzoic acid as part of their natural composition.

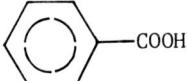

Benzoic Acid

These additives are permitted by about thirty countries throughout the world for use in a great variety of foods, particularly soft drinks. Long-term tests with rats have shown that no accumulation in the body occurs. The body excretes benzoic acid as hippuric acid within 9-15 hours of eating food containing it.

Sorbic Acid is naturally present in some fruits and is a selective growth inhibitor for certain moulds, yeasts and bacteria. It is often used in place of benzoic acid, and is added to cheeses, pickles, fish products, cordials and carbonated drinks.

$$CH_3-CH=CH-CH=CH-COOH$$

(2,4-hexadienoic acid)   Sorbic acid

Sulphur dioxide, SO_2, is unique in being a most effective inhibitor of the deterioration of dried fruits and fruit juices. It is used also as an antioxidant and antibrowning agent in wine making. Sulphur dioxide destroys the vitamin, thiamine, and so its use in major sources of this vitamin, such as meat, is not permitted.

2. Colours

Coloured substances are added to foods to make them appear more attractive. Some colours have a natural origin, such as turmeric, a yellow dye that is extracted from the root of an East Indian herb. However, most food colour additives are synthetic coal-tar dyes, many of which have been found to be carcinogenic. It is interesting to note that the list of permitted red dyes has halved in the last 30 years. Opinions as to which dyes are safe vary from country to country, and even from region to region within a country.

A typical example of a synthetic coal-tar dye is Orange B, which is often used to colour orange skins:

$$NaO_3S-\text{[naphthalene]}-N=N-\text{[pyrazole with OH]}-\text{[benzene]}-SO_3Na$$
$$O=C-OC_2H_5$$

Orange B

The coal-tar designation comes from the presence of aromatic rings. It is in these rings and in the diazo (-N=N-) link that the coal-tar dye structurally resembles many known carcinogens.

OBJECTIVES

The main objectives of this activity are:

1. to quantitatively determine the preservative content of certain foods using titrimetry or ultra-violet spectroscopy

2. to separate and identify the colour additives of foods using paper chromatography.

MATERIALS

UV spectrophotometer
burette, 25 ml
separating funnel, 100 ml
pipettes
volumetric flask, 100 ml
beakers
Whatman No. 1 chromatography paper
measuring cylinder
developing tank
hair drier
hotplate
steambath

Soft drink
jelly beans
phosphoric acid, 10%
starch iodide paper
sulphuric acid, 2M
sodium hydroxide, 2M, 0.01M
diethyl ether
ammonia solution
colour standard
butan-1-ol
ethanol
starch solution
iodine solution, 0.0010M
food dye solutions

PROCEDURE

Sulphur Dioxide in Soft Drink

A preliminary test for sulphur dioxide should be carried out on about 20 ml of degassed soft drink in a small conical flask.

1. To the soft drink, add about 1 ml of phosphoric acid and heat the mixture till boiling.

2. Hold a piece of moistened starch-iodide paper in the vapours. Sulphur dioxide is indicated by a bleaching action.

Sulphur dioxide is determined quantitatively by the redox reaction with iodine solution:

$$SO_2 + I_2 + H_2O \rightarrow SO_3 + 2I^- + 2H^+$$

3. Place 100.0 ml of degassed soft drink in a conical flask. Add about 5 ml of 2M sodium hydroxide, and stand for about ½ hour.

4. Add about 1 ml of fresh starch solution and 10 ml of 2M sulphuric acid. Titrade immediately with 0.0010M iodine solution till a permanent blue colour is formed.

5. Express the sulphur dioxide concentration as milligrams per litre of soft drink.

Benzoic Acid/Sorbic Acid in Soft Drink

Benzoic or sorbic acids are determined by measurement of ultra-violet absorbance after firstly extracting with ether.

1. Place 25 ml of degassed soft drink in a 100 ml separating funnel. Mix in a few drops of 2M hydrochloric acid.

2. Extract the mixture with three portions of diethyl ether, using volumes of 50, 25 and 25 ml. Combine the extracts, and wash twice with 15 ml portions of water.

3. Dry the ether extract by swirling with anhydrous sodium sulphate. Decant off, rinsing the residue with 20 ml of ether.

4. Evaporate the extract *almost* to dryness on a steam bath. Completely dry in a stream of air.

5. Dissolve the residue in 100 ml of 0.01M sodium hydroxide in a volumetric flask.

6. Record the ultraviolet spectrum of the solution from 190 to 270 nm against a sodium hydroxide blank. In dilute sodium hydroxide solution, benzoic acid has a maximum at 200 nm ($\varepsilon = 130\ M^{-1}\ cm^{-1}$), while sorbic acid has a maximum at 254 nm ($\varepsilon = 2.48 \times 10^4\ M^{-1}\ cm^{-1}$).

7. Calculate the concentration of either benzoic or sorbic acid in the sodium hydroxide using Beer's Law:

$$A = c\ell\varepsilon$$

where A is the absorbance, c is the concentration (M), ℓ is the curvette width (cm), and ε is the extinction coefficient ($M^{-1}\ l^{-1}$)

8. Express the preservative concentration as milligrams per litre of soft drink.

Food Colours in Jelly Beans

(Note: The colouring in soft drink may be used as an alternative).

Identification of food colours is performed by firstly extracting the dye into butanol. The extracted dye is then analyzed by paper chromatography against colour strandards.

1. Add about 25 ml of warm water to 3-6 jelly beans of each colour. Use enough jelly beans to obtain a good depth of colour. Allow to stand for about 20 minutes or till they look white. Do not let stand for too long or too much sugar will be extracted with the colours.

2. Allow the coloured solution to cool, then transfer to a 100 ml separating funnel. Add 25 ml of butanol, about 1 ml of 10% phosphoric acid, then shake. Allow the layers to separate. Run off the organic layer into a small beaker, then evaporate off the solvent on a steam bath and using a stream of air.

3. Take up each dye residue in 1 or 2 drops of 1% ammonia solution to obtain the solutions needed for chromatography.

4. Spot each colour onto the chromatography paper at about 3 cm from the bottom edge. Spot in a stream of warm air so that the spots do not get too large.

5. Spot weak solutions of reference dyes on to the same paper. Dry the paper at 105°C for a few minutes.

6. Roll the paper into a cylinderical form and staple the ends about one-third of the way in from the end. Staple the paper in such a way that the ends do not meet - otherwise the solvent will flow more rapidly at that point and form an uneven front (see Fig. 3.2.1).

7. Develop the chromatogram using the solvent system butanol:ethanol: 2% ammonia (3:1.2). Leave for 1½ hours or until the solvent has run about 15 cm.

8. After drying the paper, identify the food dyes by matching the spot colours and positions with those of the reference dyes.

QUESTIONS

1. What concentrations of sulphur dioxide, benzoic/sorbic acid were found?

2. For the foods studied, what levels of preservatives are allowed by the local food and drug regulations? Are the observed figures below these levels?

3. What food dyes were identified in the jelly beans?

4. What food dyes are permitted by the food and drug regulations? Are these food dyes water of fat soluble?

FOOD ADDITIVES AND CONTAMINANTS

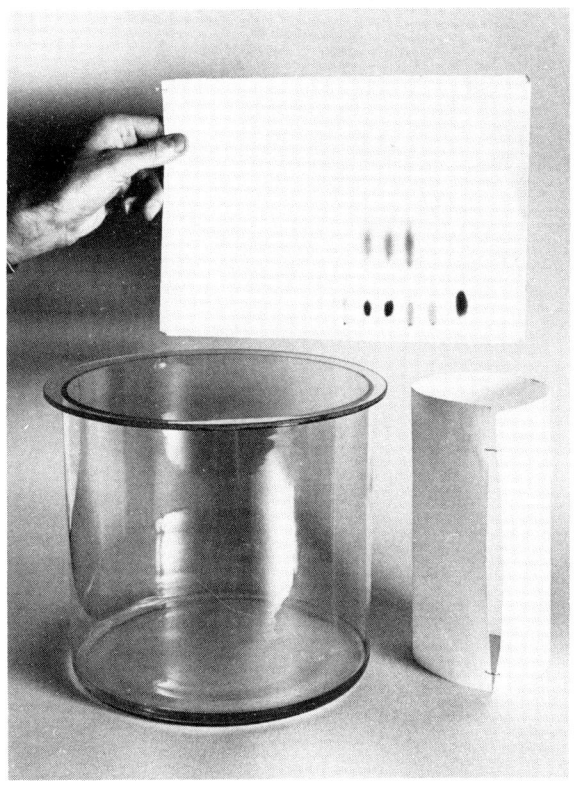

Fig. 3.2.1 Developed paper chromatogram of food dyes. Shown are a developing tank and a recommended way of rolling the paper (Photo: John Watson).

BIBLIOGRAPHY

Jones, C., Gadler, S.J. and Engstrom, P.H. (1972), *Pollution: The Food We Eat*, Lerner Publications.

Selinger, B. (1975), *Chemistry in the Market Place*, Australian National University Press, Canberra.

Anon. (1974), Toxicological Evaluation of Certain Food Additives with a Review of General Principles and of Specifications, *Technical Report Series*, No 539, World Health Organisation, Geneva.

3.3 ERUCIC ACID CONTENT OF BREAD

INTRODUCTION

Erucic acid, or 13-docosenoic acid, is a mono-unsaturated fatty acid with a 22-member carbon chain. It is found in the oil of the rapeseed plant, *Brassica napa*, where it can make up to 50% of the total fatty acid content. Erucic acid has been implicated in the causation of heart lesions and other biochemical abnormalities in experiments with laboratory animals. Because of this toxicity the use of rapeseed oil in such foods as margarine, vegetable shortenings and salad oils has been restricted.

$$CH_3(CH_2)_7CH=CH(CH_2)_{11}COOH \qquad \text{Erucic acid}$$

Because of its viscosity, stability and non-stick properties, rapeseed oil has found wide use as a releasing agent in the baking industry. The oil is applied either to the container or to the dough surface to permit easy removal of the bread after baking. However, chemical studies have shown that this practice causes contamination of the bread with erucic acid. Most of this contamination is contained in the bottom, side and end surfaces in contact with the container.

The analysis for fatty acids is routinely carried out by first transesterifying the fat (triglycerides) to form the fatty acid methyl esters. This transesterification is usually carried out by a methanolic sodium hydroxide, boron trifluoride treatment:

fat (triglyceride)   *glycerol*   *fatty acid salt*

$$\begin{array}{l} CH_2OOCR \\ | \\ CH-OOCR \\ | \\ CH_2-OOCR \end{array} \xrightarrow{CH_3OH/NaOH} \begin{array}{l} CH_2-OH \\ | \\ CH-OH \\ | \\ CH_2-OH \end{array} + \quad 3R.COO^-\;Na^+$$

$$\downarrow CH_3OH/BF_3$$

$$3R.COOCH_3$$

methyl esters

The methyl esters so produced are relatively volatile, and are well suited to gas chromatographic analysis using flame ionization detection. With this type of detection the gases emerging from the column are burnt in a small hydrogen flame between two plates at a high electric potential difference. Ions and radicals are formed in the flame, causing a small electric current to flow which is amplified and recorded. Flame ionization detection is very sensitive for C-C and C-H containing compounds.

OBJECTIVES

The main objectives are:

1. to extract and transesterify the fatty acids in some bread samples.

2. to identify and quantify some of the fatty acids using a flame-ionization gas chromatograph.

MATERIALS

Flame-ionization gas chromatograph
mortar and pestle
separating funnels, 250 ml
beakers, 100 ml
graduated tubes, 50 ml
measuring cylinders, 50 and 10 ml
filter funnel

Chloroform
methanol
0.5 M methanolic sodium hydroxide
  solution
14% (w/v) of BF_3 in methanol
hexane
saturated brine
anhydrous sodium sulphate
  crystals
standards of erucic acid, palmitic
  acid and oleis acid, 100 µg/ml
  in hexane, as the methyl esters.

PROCEDURE

The analysis normally require two laboratory sessions. The first session can be devoted to separation of the lipids, transesterification, and extraction of the methyl esters. A selection of samples (2 - 3) should be chosen for analysis. The procedure outlined applies to bread samples. However, other samples such as margarine and salad oil may be similarly treated, omitting the lipid separation steps 1-3.

1. Weigh a suitable portion of bread. Dry in an air oven at $103^\circ C$, then grind to a powder in a mortar and pestle.

2. Weigh out duplicate portions (about 20 g) of the ground bread. Grind each portion to a thin paste with 30 ml of chloroform/methanol solution (2 : 1, v/v).

3. Filter the slurry through a tight plug of cotton wool into a small beaker, using chloroform to complete the rinsing. Warm the filtrate under a stream of dry air to remove the solvent.

4. Add 5 ml of 0.5 M methanolic sodium hydroxide solution too the lipid extract, and boil gently on a steam bath for 3 minutes. Cool, add 5 ml of 14% BF_3 methanol solution, then boil for a further 2 minutes.

5. Transfer the crude solution of methyl esters to a 250 ml separating funnel. Add 30 ml of hexane; 20 ml of saturated brine, and shake vigorously. Leave stand, then run off the aqueous layer. Wash the hexane once with 30 ml of water.

6. Run off the organic layer into a small beaker containing some anhydrous sodium sulphate then, after thorough mixing, decant the liquid into another beaker. Evaporate to about 5 ml, transfer to a graduated tube, then complete the evaporation. Make up the residue to 5 ml with hexane and add a few grams of anhydrous sodium sulphate.

7. Set up the gas chromatograph using the following conditions:

   2 m x 3 mm column packed with 0.3% OV-101 on 100 - 120 Chromosorb WAW
   column temperature $230^\circ C$

inlet and detector temperatures, 245°C
and detector temperature, 245°C
carrier gas, N_2 at 30 ml/min.

8. Use a syringe to inject 0.5 ml of hexane into the vial containing the methyl ester residues. Shake to dissolve. Chromatograph 1 μl volumes of this sample solution.

9. Inject 1 μl volumes of the erucic acid, palmitic acid and oleic acid methyl ester standards. Identify the corresponding peaks in the sample chromatogram. Calculate the concentrations of these acids and express as milligrams per gram of bread or oil.

QUESTIONS

1. What are the erucic acid concentrations of your bread or oil samples?

2. What standards or recommendations cover erucic acid in foods? How do your results compare with these?

3. What are some other commonly occurring C_{22} fatty acids with one double bond?

BIBLIOGRAPHY

1. Christophersen, B.O., Svaar, H., Langmark, F.T., Gumpen, S.A., and Norum, K.R., *Rapeseed Oil and Hydrogenated Marine* Oils *in Nutririon*, Ambio, *5*, 169 (1976).

Wills, R.H.B., Wootton, M. and Hopkirk, G. Erucic Acid, an Accidental Additive in Bread, *Nature, 263,* 504 (1976).

3.4 DDT IN HUMAN MILK

INTRODUCTION

The metabolism of DDT in a variety of organisms has been extensively investigated, particularly with insects. However, the fate and effects of DDT with man are not so well understood and are the subject of considerable debate. It is well known that the immediate lethal toxicity of DDT with man is low but the possible longer term effects have not been intensively studied.

The information available indicates that DDT administered orally accumulates in bile, brain, blood and the liver and is slowly eliminated in the urine and faeces as principally DDA (bis-(p-chlorophenyl)acetic acid). A simplified portion of the degradation pattern of DDT in mammals is shown in Fig. 3.4.1

DDT and its metabolites also tend to accumulate in the fatty tissues of animals. In addition a number of investigations have shown that these substances bio-magnify, i.e. they exhibit increasing concentrations in a food chain in relation to trophic level. Man is an omnivore but exhibits many of the characteristics of a top carnivore. It therefore could be expected that the fatty tissues of man would often exhibit comparatively high concentrations of DDT and metabolites (see Tables 3.4.1 and 3.4.2).

TABLE 3.4.1   DDT Residues in the Fatty Tissues of Man

COUNTRY	DDT (ppm)
India (Delhi)	26
Israel	19
Hungary	12
USA	12
Czechoslovakia	10
Canada	6
France	5
Italy	5
Denmark	3
United Kingdom	3
West Germany	2

(From: *Third Report of the Research Committee on Toxic Chemicals*, Agricultural Research Council, London, 1970; there is some evidence to suggest that these concentrations may have reduced in some cases in recent years).

A common technique used in analysing for chlorinated hydrocarbon insecticides involves the use of gas chromatography (see Section 2.2) with the electron capture detector. The electron capture detector is very sensitive to the chlorinated hydrocarbons and can detect them in very low concentrations. In principle the detector has a small radioactive electron source which produces a standing current across the effluent gases from the gas chromatography column. When the carrier gas only emerges, a constant signal is produced and is recorded as a straight line on the recorder chart, but when another substance is contained in the carrier gas some of the electrons may be absorved, resulting in a reduction in the current produced and giving a corresponding signal on the recorder, Halogenated compounds are very efficient at absorbing electrons and so give large signals compared to other substances. Fig. 3.4.2 is an example chromatogram of mother's milk extract.

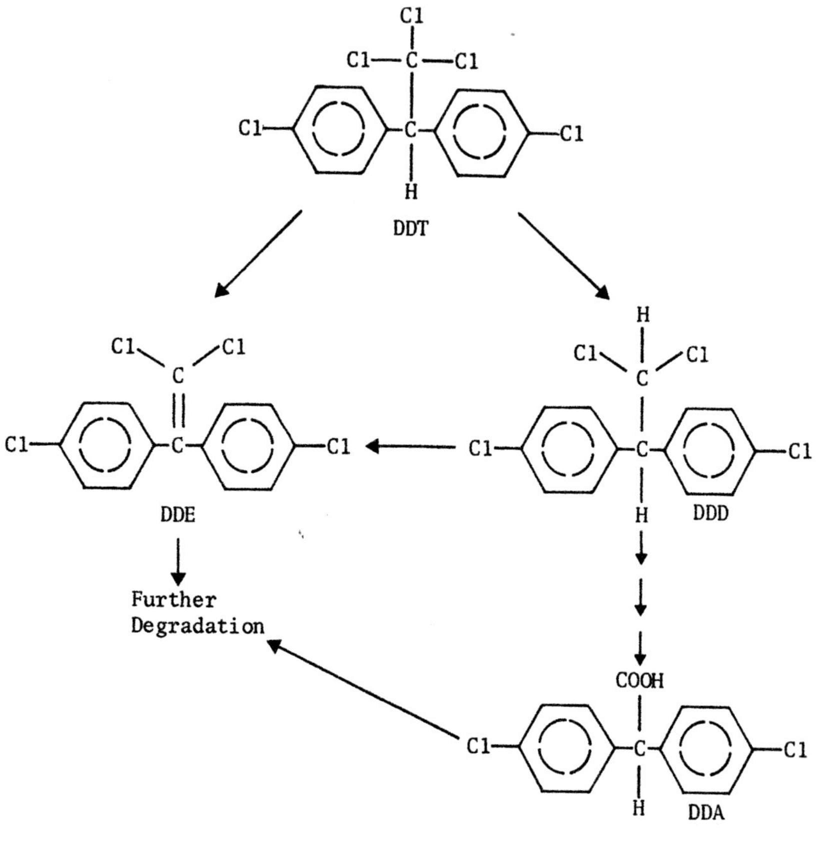

Fig. 3.4.1 Simplified Pattern of the Metabolism of DDT in Mammals.

TABLE 3.4.2 Average Levels of Organochlorine
Residues in Human Milk of Swedish Women, 1976-1977

	μg/kg whole milk	mg/kg, milk fat
DDT	10	0.3%
DDE	43	1.49
Dieldrin	0.7	0.025
Hexachlorobenzene	3.3	0.11
PCB	.30	0.93

(Adapted from G. Westoo and K. Noren, Organochlorine Contaminants in Human Milk, Stockholm, 1966-1977. *Ambio, 7* 62 (1978).

OBJECTIVES

The main objectives of this activity are:

1. to isolate and purify the DDT residues in human milk
2. to identify and quantify the pesticide residues by gas chromatography with the electron capture detector.

MATERIALS

Electron capture gas chromatograph
Rohrig extraction tube
separating funnel, 250 ml
chromatography column, 40 x 2 cm
beakers
centrifuge tubes
pipette, 10 ml
graduated cylinder, 25 ml

Potassium oxalate
methanol
diethylether
petroleum spirit, 60-75°C
hexane
anhydrous sodium sulphate
Florisil, deactivated 5% (w/w) with water
standard solutions of DDT, DDE, dieldrin and dichlorvos

PROCEDURE

The analysis normally requires two laboratory sessions. The first can be devoted to the separation and purification of the pesticide residues, while the second can be used for the chromatographic analysis.

Isolation of Pesticide Residues

The first part of the residue separation requires extraction of the contaminated milk fat from the whole milk sample.

1. Pipette 10 ml of mothers milk into a Rohrig extraction tube. Add about 1 g of potassium oxalate and shake the contents of the tube. Add 1 g of potassium oxalate to a second tube and carry this through the analysis to assess the system blank.

2. Mix in 10 ml of methanol. Add 25 ml of diethyl ether and shake vigorously to extract the fat. Add a further 25 ml of petroleum spirit and shake again.

3. After settling, decant the ether-petrol layer into a 250 ml separating funnel containing 60 ml of deionized water.

4. Repeat the diethyl ether-petroleum spirit extraction twice more using 25 ml of each solvent each time. Add the new extracts to the separating funnel.

5. Shake the combined extracts in the separating funnel, then run off the water layer. Wash the remaining organic layer twice with deionized water, discarding the washings.

6. Run off the organic layer into a dry beaker containing some anhydrous sodium sulphate. Swirl, and decant the liquid into another, tared beaker.

7. Concentrate and dry the extract under a stream of clean dry air. Weigh the residual milk fat concentrate.

The insecticide residue is now contained in the milk fat concentrate from which it can be separated by column chromatography:

8. Sorb the milk fat concentrate onto a 2 x 25 cm column of Florisil (deactivated 5% w/w with water) topped with 1 cm of anhydrous sodium sulphate. Use a minimum volume of hexane to transfer the concentrate and do not allow air to enter the column packing.

9. Elute twice with 50 ml portions of 6% diethyl ether-petroleum spirit. Collect the combined effluents in a beaker and evaporate to a small volume in a stream of dry air. Continue the evaporation in a centrifuge tube.

Chromatographic Analysis

The final part of the analysis requires gas chromatography. For this, an electron-capture gas chromatograph needs to be set up with the necessary column, gas flow and temperatures. A suitable set of conditions are as follows:

4 mm x 2 m column of 1.5% OV.17/1.95% SP-2401 on 100-120 Chromosorb WAW
column temperature, 175-183°C
detector and injection port, 213°C
carrier gas nitrogen at 30 ml min^{-1}.

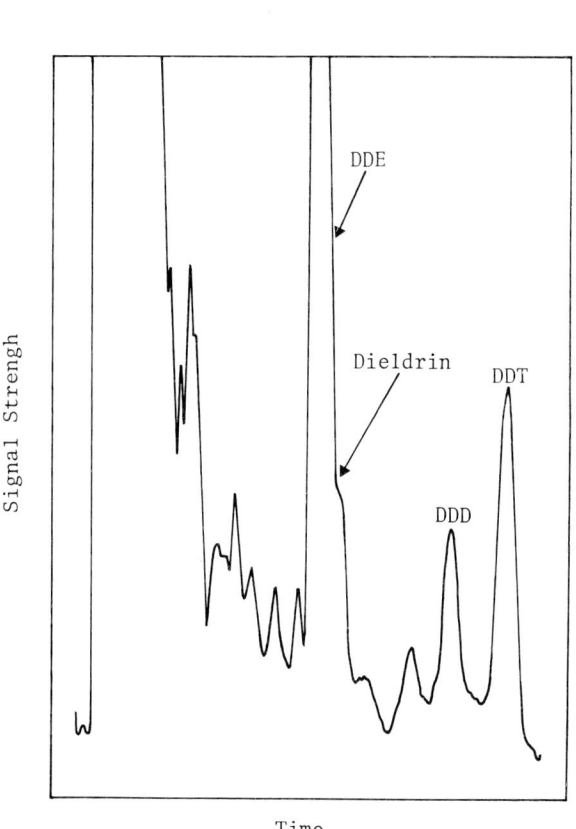

Fig. 3.4.2 Electron-capture gas chromatogram of mother's milk extract. Note that most of the DDT has metabolised to DDE and DDD. Often the total DDT content (DDT & metabolites) of samples is quoted in the literature.

1. Make up the pesticide extract to 2.0 ml with hexane. Inject 1 μl into the gas chromatograph and record the chromatogram.

2. Chromatograph 1 μl volumes of DDT, DDD, DDE, dieldrin and dichlorvos standards. Use concentrations so that the peaks are roughly the same size as those for the extract.

3. Identify the pesticide residues in the human milk. Calculate their concentrations, and express the values as microgram *per* gram of milk fat extract.

QUESTIONS

1. What are the quantities and identity of chlorinated hydrocarbon insecticides in the sample examined?

2. How does the pesticide content of the human milk fat compare with the tolerances set for cows milk set by the state or national food and drug regulations?

3. What is the estimated daily pesticide intake of babies fed on breast milk? How does this compare with the acceptable daily intake set by the WHO?

4. Comment on factors likely to effect the levels of DDT in the sample you have examined.

BIBLIOGRAPHY

Bradt, P.T. and Herrenkohl, R.C., DDT in Milk: What Determines the Levels. *Sci. Total Environ.*, *6*, 161 (1976).

World Health Organization, Pesticide Residues in Food, *Wrld. Hlth. Org. Tech. Rep. Series*, No. 474 (1971).

PART 4

Chemical Ecology

4.1 CHEMICAL DEFENCE OF THE MONARCH BUTTERFLY

INTRODUCTION

Ecology is the study of the interactions of animals and plants and how they relate to their abiotic environment. Chemical ecology is that particular branch of ecology which is concerned with the study of how animals and plants interact with their chemical environment. Such interactions can take the form of the release or secretion of substances in natural associations of plants and animals which influence the behaviour, growth, other characteristics of other organisms in the group.

Plants have evolved a wide variety of chemicals that influence other organisms. Those substances which affect other plants are termed allelochemicals. Examples of allelochemicals are juglone, produced by certain eucalyptus species, and cineole, a terpenoid released from leaves of the Californian mint or sagebrush. Both of these chemicals inhibit the growth of other plants in the immediate vicinity of the parent plant, so reducing competition for nutrients and sunlight.

Cineole   Juglone

Plants also produce substances that influence the behaviour of insects. These substances include alkaloids, tannins, glycosides, terpenes, and organic acids. Such substances do not appear to play any metabolic role in the plant and are termed secondary plant substances. The role of these secondary plant substances is varied, ranging from insect attraction in the case of volatile compounds emitted by flowers to defense in the case of compounds which are toxic to foraging animals.

Insects have developed chemicals for a variety of purposes including defense from predators and communication with other insects. Defense chemicals often have irritant properties to deter potential predators, and include organic acids, aldehydes, quinones and terpenes. For example, benzoquinone and related substances are used by beetles of the family *Tenebrionidae* (darkling beetles). These beetles have a gland in their tail which they can extrude covered with the repellent compounds. Chemicals which permit communication between animals of the same species are termed pheremones. Some pheremone chemicals are 6,8- nonadecadiene (sex attractant, gypsy moth), formic acid (alarm substance ants) and phenylacetic acid, (trail marking, mongolian gerbil)

$CH_3(CH_2)_4CH=CH-CH=CH-(CH_2)_9CH_3$

6,8-nonadecadiene

HCOOH

Formic acid

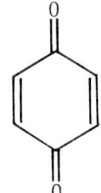

Benzoquinone

$\langle\bigcirc\rangle-CH_2-COOH$

Phenylacetic Acid

Secondary plant substances have sometimes played an important role in the coevolution of plant and animal species. A well-known example is the animal-plant system composed of the monarch butterfly, *Danaus plexippus*, and plants of the genus *Asclepias*, commonly known as milkweeds. The *Asclepias* plants produce a complex of steriods known as cardiac glycosides. These substances are vertebrate toxins and confer a certain immunity to the plant from herbivores. However, at some time in the past, the monarch butterfly evolved the ability to use *Asclepias* plants as hosts for its larvae. The feeding larvae store and concentrate these toxins, so that they and the subsequent adult butterfly are conferred with protection from predators such as birds. The cardiac glycosides have a powerful emetic effect on birds, so that a bird feeding on the insect will soon vomit. Birds apparently soon learn to recognize the brightly coloured butterfly and its larvae as unsatisfactory prey.

The cardiac glycosides are so named because of their action on heart muscle. They belong to group of C_{23} steroids known as cardenolides:

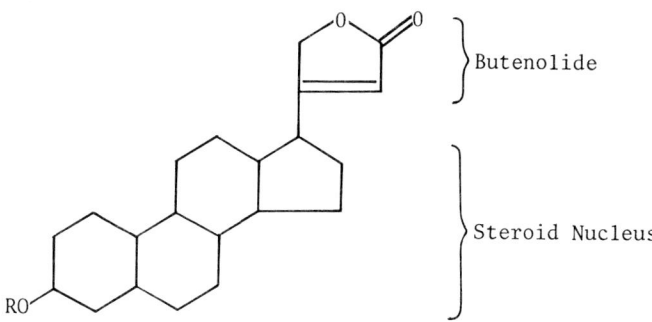

a Cardenolide

The cardenolides are characterized by the presence of a five-membered butenolide ring attached to the steroid nucleus, while other substituents around the nucleus contribute to the variation in the compounds. In nature, the cardenolides most commonly occur as glycosides where the substituent R is one or a series of linked sugar groups. Plants usually produce mixtures of these glycosides, two examples of which are ouabain and digitoxin:

Digitoxin

Ouabain

OBJECTIVES

The main objectives are:

1. To determine the total cardiac glycoside content of Monarch larvae, and their host plant, *Asclepias* spp.

2. To observe that the cardiac glycoside content of the lavae is derived from the food plant and increases with larval growth.

MATERIALS

beakers
buchner funnel
sonifier or mortar and pestle
separating funnels
pipettes
spectrophotometer

monarch lavae
Asclepias leaves
chloroform
methanol
ethanol
petroleum spirit
diethyl ether
ouabain
sodium hydroxide solution, 1M
sodium carbonate
  solution, 2M
3,5-dinitrobenzoic acid

PROCEDURE

Samples of Monarch larvae and milkweed plant may be collected during the warmer periods of the year in North and South America, Australia and South-East Asia. The larvae are about 5-50 mm long and have prominent white, yellow and black lateral stripes (Fig. 4.1.1) About 20 of the larger individuals representing those in the final or 5th instar should be collected along with 50-100 of the younger larvae. Young larvae tend to inhabit the young, growing shoots of the milkweed while the older larvae prefer the larger, more mature leaves. Pupae or adult butterflies may be collected in place of the larger larvae.

Both plants and larvae are treated by solvent extraction to remove the cardiac glycosides. Plant extracts require some cleanup by solvent partition to remove interfering pigments.

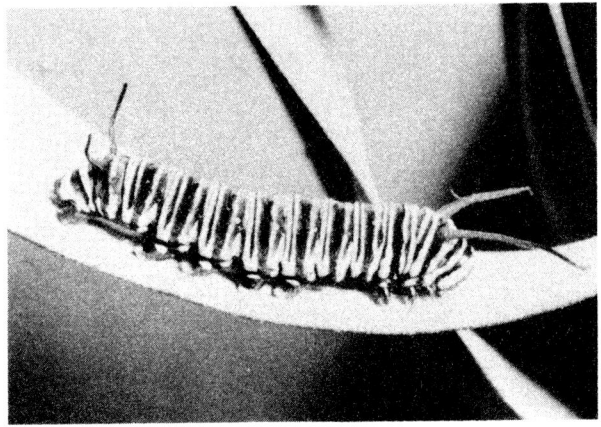

Fig. 4.1.1 Lava of the monarch butterfly feeding on leaves of milkweed plant (Photo: Dudley Nott).

1. Divide the larvae into two groups. One group should contain 20 of the larger larvae, preferably in the fifth instar if this can be recognized. The other group should contain smaller individuals, <35 mm, in earlier instars and in sufficient number so that the total weight is 10-20 g. Weigh both groups of lavae.

2. Crush each group in 30 ml of chloroform-methanol (2:1) in a 100 ml beaker and heat till boiling. Cool, then filter the mixture through a small Buchner funnel, rinsing with 10 ml of solvent.

3. Evaporate the filtrate under a stream of air with warming. Dissolve the residue in 10 ml of ethanol and store till the colourimetric assay.

4. Weigh out about 200 g of fresh *Asclepias* leaves and air dry at 60°C for 48 hours. Grind to << 0.5 mm using either a mill or a mortar and pestle. Weigh 5 g of the powdered leaf and mix with 100 ml of ethanol-water (2:1). Sonify for 10 minutes with an ultrasonic cell disrupter. If a sonifier is not available, mascerate with a mortar and pestle and let the

mixture stand for 24 hours. Filter through a fast filter paper, rinsing the beaker and residue with 50 ml of ethanol-water. Discard the residue.

5. Extract the filtrate three times with 50 ml volumes of petroleum spirit in a 250 ml separating funnel. The petroleum layers contain fatty substances and are discarded. Evaporate the aqueous alcohol layer to about 20 ml under a stream of dry nitrogen with warming.

6. Extract the concentrate in a 100 ml separating funnel with two volumes each of 50 ml ether, 40 ml chloroform, and 30 ml chloroform-ethanol (3:2). Keep the extracts separate. Wash each extract twice with 50 ml volumes of 2 M Na_2CO_3 solution then twice with 50 ml of water to remove acids.

7. Evaporate the organic extracts under nitrogen with warming. Dissolve each extract in 10 ml of ethanol and store till the colourimetric assay.

The colourimetric determination is based on the reaction of the active methylene group in the butenolide ring. Roughly the same colour intensity is given by various cardiac glycosides, so that the method allows for quantitative estimation of the gross cardiac glycoside content. A commercially available glycoside such as ouabain or digitoxin can be used as a standard.

8. Place 1.5 ml of the ethanolic sample solution in a small test tube. Add 1 ml of 2% 3,5-dinitrobenzoic acid in ethanol and 0.5 ml of 1M aqueous sodium hydroxide solution. Quickly mix the solutions and transfer to a 1 cm cuvette. Read the absorbance at 565mm against a reference mixture containing ethanol in place of the 3,5-dinitrobenzoic acid solution. The pink colour that develops is not stable and fades after about 4 minutes. Record the maximum absorbance reached in this time.

9. Prepare a standard solution of 1 mg/ml of ouabain in ethanol. Pipette 0.2, 0.4, 0.6 and 0.8 ml volumes into a series of test tubes and make up to 1.5 ml with ethanol. Treat each with the colourimetric agent as above and measure the absorbances against a reference containing ethanol in place of ouabain solution. Plot the absorbances against mg of ouabain to obtain a straight line passing through the origin.

10. Read off from the standard curve the cardiac glycoside content of the sample solutions. Calculate the glycoside concentration of the lavae or leaf samples using the equation:

c.g. concentration (mg/g as ouabain) = mg ouabain per 1.5ml $\times \dfrac{10}{1.5} \times \dfrac{1}{\text{mass of sample, (g)}}$

QUESTIONS

1. What concentrations of cardiac glycosides were found? Are there any systematic relationships related to the ecology of the insects?

2. Other plants, besides *Asclepius*, produce cardiac glycosides. What are these plants, and what major toxins do they produce? Do they have any ecological role?

BIBLIOGRAPHY

Dixon, C.A., Erickson, J.M., Kellett, D.N., and Rothschild, M., Some Adoptations between *Danaus plexippus* and its Food Plant, with Notes on *Danaus chrysippus* and *Euploea core* (Insecta:Lepidoptera), *J. Zool., Lond.* 185, 437 (1978).

Urquhart, E.A. (1960), *The Monarch Butterfly*, University of Toronto Press, Toronto.

PART 5

Field Survey

5.1 STREAM POLLUTION

INTRODUCTION

Most major cities contain a limited number of large waterways such as bays, harbours, and rivers together with a network of small urban streams. During the growth of the cities, many streams have been enclosed or otherwise converted into drains. In addition, these streams have been subject to an increasing pollution load from contaminated urban run-off water, industrial operations, and other activities.

In recent years ther has been increasing interest in urban streams with the streams and adjacent land being used as parks and natural recreation areas by city dwellers. Polluted streams represent a health risk and also contaminate the larger bays and waterways, reducing their value for fishing, swimming and aquatic sports.

An investigation of stream pollution should include a survey where various physical and chemical characteristics of the stream are measured. These measurements may then be used to assess if the water quality meets certain criteria for human use and for the maintenance of aquatic life. As well, the survey should look for any changes in these characteristics over time or place, so that possible links can be made with influencing factors such as water flow, climate, surrounding land use, and the input of effluents and wastes.

OBJECTIVES

The main objectives are:

1. to evaluate the quality of the water of an urban creek in terms of requirements for the maintenance of aquatic life

2. to identify and evaluate natural and man-made factors influencing water quality over the total length of the creek.

SURVEY PLANNING

Mapping

Aerial and topographical maps of the stream and its catchment should be obtained. From these maps, major features of land use and sources of point discharges such as factories and sewage plants can be identified. A series of test and sampling sites should be selected so that they are connected with these features; in particular where point discharges are known or suspected, sites up- and downstream should be selected in order to assess the effect of these discharges on water quality. The sites should be checked for ease of access and suitability.

Timing

Long-term water quality surveys are not appropriate for small groups of students. Because water flow and water use change with time, a rapid survey

should be planned to assess the water quality along a stream over a limited
time period. Also, many streams are tidally influenced for many kilometres,
and in these cases the survey should be conducted at a fixed point in the
tidal cycle. A factor to be taken into account with tidal streams is the
tidal "lag" or the time difference in the tidal cycle between the mouth and
a place upstream. Because of this lag, it is often more practicable to begin
a rapid survey at the stream's mouth.

Field and Laboratory Tests

Twelve tests are suggested to be included in the survey. The list is not
exclusive, but covers parameters which have an important bearing on water
quality. Furthermore, the tests are relatively straightforward, and can be
attempted by small groups of students on large numbers of water samples within
a reasonable time.

At least four tests should be carried out at each site in the field survey.
These tests are:

1. Temperature measurement
2. Dissolved oxygen measurement
3. Conductance and/or salinity measurement
4. pH measurement

In addition, more complex field tests and laboratory analysis of field samples
may include the measurement of:

5. Phosphorus content
6. Biochemical oxygen demand
7. Chlorophyll-a
8. Suspended solids
9. Oil and grease in sediments
10. Ammonia
11. Available chlorine
12. Streamflow and discharge

SAMPLE COLLECTION AND STORAGE

Only in some instances will it be possible to perform a measurement *in situ*,
and many tests require a water sample to be taken. Standard practice is to
take a measurement or a water sample from about 15 cm below the surface in
mid-stream. Large, floating material should not be included in the sample. The
collected sample may either be tested on site or stored for later laboratory
analysis. Table 5.1.1 summarizes some procedures for sampling and storage.
In general, a sample is used to rinse and fill a precleaned container, which
is then sealed and cooled to about 4°C in a portable ice-box. Glass and
plastic containers may be used, but sterile, disposable, sealable plastic
bags* are increasingly popular.

CAUTION: *Students should exercise care in handling polluted waters as these
are often a health hazard.*

---

* Whirl-paks^{TM}, or their equivalent.

TABLE 5.1.1 Recommendations for Sampling and Preservation of Samples

Measurement	Vol. Req. (ml)[a]	Preservative	Maximum holding time
Temperature	1000	Det. *in situ*[b]	nil
Salinity	100	Cool, $4°C$ Det. *in situ*[b]	24 hr.
pH	25	Cool, $4°C$ Det. *in situ*[b]	6 hr. nil
Dissolved oxygen	–	Det. *in situ*[b]	nil
BOD	1000	Cool, $4°C$	6 hr.
Turbidity	100	Cool, $4°C$ Det. *in situ*[b]	7 days nil
Residues	250	Cool, $4°C$	7 days
Phosphorus	50	Cool, $4°C$	7 days
Ammonia	400	Cool, $4°C$ H_2SO_4 to pH < 2	24 hr.
Available chlorine	1000	Cool, $4°C$	<6 hr.
Chlorophyll	350	Cool, $4°C$	6 hr.
Oil and Grease	1000	Cool, $4°C$ H_2SO_4 to pH < 2	24 hr.

[a] This is the minimum recommended volume for one determination.

[b] Samples cannot be held and the measurements must be made near the site if not *in situ*.

FIELD AND LABORATORY TESTS

Temperature, Dissolved Oxygen, Conductance or Salinity, and pH

Some commercial instruments are available that measure several or all of the above parameters. These parameters should be measured *in situ*.

Measurements of conductance are usually expressed in micro- or millimho per centimetre, µ/cm or m/cm (the equivalent S.I. units are micro- and milliseimens, µS and mS). Some instruments also have a salinity scale expressed in parts per thousand (ppt); this scale may be used for estuarine or seawaters with salinities greater than 2ppt, and usually is not applicable to freshwaters. Values in conductance units may be converted to salinity units (>2ppt) using Fig. 5.1.1.

Dissolved oxygen meters are normally calibrated against air-saturated fresh water. A correction then has to be made for measurements in saline water. Some instruments have this facility built-in, but, where this is absent, correct the saline water value using the equation:

$$D.O. \text{ (corr)} = \frac{(D.O. \text{ of saline water air saturated at measurement temp})}{(D.O. \text{ of fresh water air saturated at measurement temp})} \times \text{meter reading of saline water sample}$$

The dissolved oxygen values of air saturated saline and fresh waters are given in Table 1.1.1.

Phosphorus, Biochemical Oxygen Demand, and Chlorophyll-a

Essentially the procedures described in Sections 1.1 and 1.2 apply here. For orthophosphate and chlorophyll-a analysis, it is advisable to filter the water sample on site: the filtrate is stored for orthophosphate analysis, and the filter is stored with 10 ml of 90% acetone-water in a glass vial for chlorophyll-a analysis.

Suspended Solids

Materials. Membrane filter kit, GF/C glass-fibre filter disc, measuring cyclinder, dessicator.

Procedure. 1. Rinse several glass-fibre filter discs with distilled water using a membrane filter apparatus. Dry the discs at $105°C$ for 1 hr. Allow to cool for 2 hr in a dessicator containing silica-gel, then weigh each filter disc.

2. Filter 100-500 ml of water sample through a glass-fibre disc, recording the volume used. Rinse with distilled water. Redry the disc at $105°C$ for 1 hr, cool in a dessicator, and reweigh.

Oil and Grease in Sediments

This determination covers petroleum ether extractable materials such as

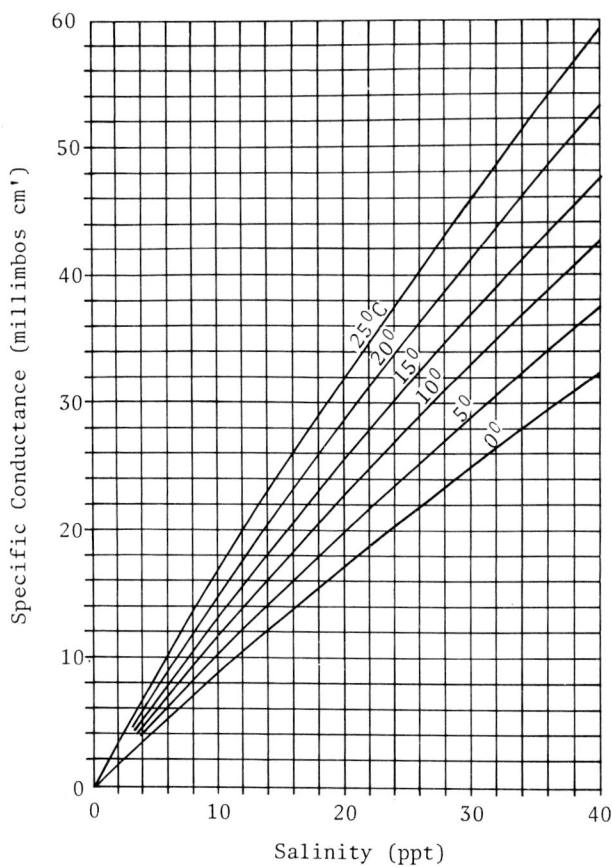

Fig. 5.1.1 Conversion between specific conductance (millimho/cm) and salinity (parts per thousand) for diluted seawaters at several temperatures.

non-volatile hydrocarbons, oils, fats, waxes, soaps and similar materials. It is not applicable to hydrocarbons that volatilize below 70°C.

Sampling   Sediment is collected with either a benthic grab or a PVC tube pushed into the sediment. About 500g is transferred to a weighed, petroleum rinsed jar. About 5 ml of 4 M sulphuric acid are added to prevent deterioration.

Materials. Wrist shaker, 250 ml separating funnel, steambath, 250 ml conical flasks, dessicator, petroleum ether, 4 M sulphuric acid, siliconized filter paper.

Procedure.   1. Reweigh the sample jar to obtain the weight of sediment. Add sufficient distilled water to form a slurry, shake, and check with pH paper that the mixture has pH <2. Add sulphuric acid if necessary.

2.   Add 100 ml of petroleum ether to the mixture. Shake on a mechanical shaker for 10 min, then decant off the organic layer into a 250 ml separating funnel. Repeat the extraction with a further 50 ml of petroleum ether.

3.   Add 100 ml of distilled water to the combined extracts and shake. Run off and discard to lower agueous layer. Rewash with a further portion of water.

4.   Weigh a 250 ml conical flask (predried in a dessicator for 2 hr). Filter the organic extracts into the flask through a siliconized filter paper. Evaporate off the solvent on a 70°C steambath with the air of a stream of air. Cool the flask for 2 hr in a dessicator and reweigh. Express the residue mass as mg per kg of wet sediment.

Ammonia

Ammonia can be determined potentiometrically using a selective ion electrode and a millivoltmeter. In this method, the sample is made basic to convert ammonium ions to dissolved ammonia which is in turn detected by the gas-sensing specific ion electrods.

Materials.   Ammonia electrode and millivolt meter, magnetic stirrer, 250 ml beaker, pipettes, 10 M sodium hydroxide, 1000 mg/l N solution (3.60 g of ammonium chloride per litre of distilled water).

Procedure.   1. Prepare a series of standard solutions containing 0.1-1.0 mg/l N by diluting the stock ammonium chloride solution with distilled water.

2.   Place 100 ml of the lowest concentration standard in a 250 ml beaker. Insert the electrode and stir the solution. Add 1 ml of 10 M sodium hydroxide solution. Check pH paper that the mixture has a pH >11. Within three minutes the reading should reach a maximum and then decrease Record the maximum value.

3.   Repeat this procedure with the remaining standards, going from the lowest to the highest concentration. Use semi-logarithmic graph paper to plot the ammonia concentration in mg of N per litre on the log axis *versus* electrode potential in mV on the linear axis. A straight line should be obtained.

4. Follow the procedure of 2 using 100 ml of sample. Read off the ammonia concentration of the sample from the standard curve.

Available Chlorine

The determination based on the action of chlorine on potassium iodine solution which liberates iodine, ($Cl_2 + 2I^- \rightarrow 2Cl^- + I_2$) and the titration of the released iodine with standard sodium thiosulphate solution ($S_2O_3^- + I_2 + H_2O \rightarrow S_2O_3^- + 2I^- + 2H^+$).

Materials. Burette, 1000 ml conical flask, 0.1M sodium thiosulphate solution (25.5g $Na_2S_2O_3$, $5H_2O$ in 1 litre distilled water + 1 ml $CHCl_3$), potassium iodide, starch solution, glacial acetic acid.

Procedure. 1. Dilute the 0.1 M sodium thiosulphate solution to 0.01 M with distilled water.

2. Place 900 ml of water sample into a 1 litre conical flask. Add about 5 ml of glacial acetic acid and about 1 g of potassium iodide. Stir, and titrate with 0.01 M thiosulphate solution till the brown colour of iodine is almost discharged. Add 1 ml of starch solution and titrate till the blue colour is discharged.

3. Run a blank titration using 900 ml of distilled water in place of the water sample. Correct the titre for the blank, and calculate the available chlorine content of the water sample using the equation.

$$Cl \text{ (mg/l)} = \frac{\text{Corr. titre} \times 35.45}{900} \times \frac{1000}{1}$$

Streamflow

The velocity of a stream is the rate of water movement past a given point. The flow or *discharge* of the stream is the total amount of water transported past a given point in unit time, and is usually expressed in cubic metres per second.

An approximate value for the discharge of a stream may be calculated by the velocity-area method. In this method, the discharge is given by the product of the cross-sectional area of the stream and the average water velocity. The area is determined by measuring the cross-sectional profile of the stream channel, and a rough estimate of the velocity is taken as 0.8 times the mid-stream surface velocity.

The point velocity of water will vary from bank to bank and from surface to bottom. A more accurate way of applying the velocity-area method involves integration of the velocity over the flow section. Referring to Fig. 5.1.2, the cross section is divided into several segments. The average velocity of each segment is taken as roughly the measured velocity at $5/10^{th}$ the depth of water in the middle of the seament. The discharge of each segment can then be determined, and the sum of these discharges taken as the flow of the stream.

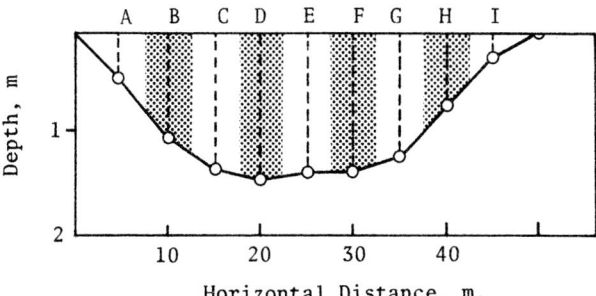

Fig. 5.1.2 Profile of a stream obtained by depth measurements at several points. Flow measurements at 0.6 depth are taken at each point. Segments (shaded and unshaded) can then be constructed in order to estimate the discharge of the stream.

Materials. Propeller flowmeter and counter, surveying staff, weighted graduated lines, measuring tape.

Procedure. 1. Stretch a measuring tape between the banks of the stream. At every 1-2m, whichever is more appropriate for the width of the stream, estimate the depth of the water using either a surveying staff of a graduated line. At each point, use a flowmeter to measure the water velocity in metres per second at 0.6 depth above the bed of the stream.

2. Draw up a cross-sectional profile of the stream on squared paper. Connect the points showing the depth of the stream bed with straight lines. Divide the cross section into segments as shown in Fig. 5.1.2.

3. Estimate the area of each segment in square metres by counting squares in the diagram. Multiply this figure by the 0.6 depth velocity to obtain the discharge of each segment. Total the discharges of the segments to obtain the flow of the stream in cubic metres per second.

PRESENTATION OF THE REPORT

Although there is no fixed format the following items should be included in the report:

1. *Table of Contents*. This should be explicit and brief.

2. *Introduction*. In this you should describe the reasons for the study, incorporating at least:

(a) previous similar studies
(b) the nature of the study area
(c) the nature of the water quality parameters studied
(d) the aims of the study.

3. *Results*. This section should summarize numerical results and include observations. The result should be presented in tables and/or graphs. Sufficient text should be included so as to identify the results. Notes should be made on the precision of the results as well as on any significant variations that can be connected with the methods used.

4. *Discussion*. This should cover the significance of your findings in relation to water quality and the use made of the stream. The discussion may include:

(a) a sketch map of the stream showing the sampling sites and adjacent major developments
(b) graphs showing the variation in the water quality parameters along the stream's length

(c) comparisons against recommended water quality criteria
(d) any relationships between the parameters
(e) identification of some factors that influence water quality
(f) estimates of the diluting capacity of the stream.

5. *Conclusions*. This section may include general comments on the standard of water quality, and guidelines for future investigations and water management.

6. *Experimental Procedures*. These should be identified and very briefly outlined. Details should only be given in the cases of procedures not described elsewhere or where there are substantial modification. Known procedures should be referenced.

7. *References*.. References to personal communications and published material should be indicated in the text. These references should be collected at the end of the report in the order in which they appeared in the text or alphabetical order of authors.

BIBLIOGRAPHY

American Public Health Association, American Waterworks Association and the Water Pollution Control Federation (1975), *Standard Methods for the Examination of Water and Wastewater*, 14th Edition, APHA, Washington.

U.S. Environmental Protection Agency (1976), *Methods for the Chemical Analysis of Water and Wastes*, U.S.E.P.A. Technology Transfer EPA-625-/6-74-003a, Washington.

SUBJECT INDEX

Aflatoxins 64
  estimation in peanuts 66
  regulatory limits 65
  TLC 67
Alkanes 36
Alkenes 37
Alkynes 37
Ammonia in water 14, 97
Anoxic conditions in water 15
Aquatic plants
  effect on dissolved oxygen 13
  effect on pH 13
Aromatic hydrocarbons 37
Atmospheric pollutants 54
  concentrations in air 54
  criteria 56
Atomic absorption spectroscopy 52

Benzoic acid in soft drink 71
Bioassay 45
  analysis of data 17
Biochemical oxygen demand
    (BOD) 4, 8, 10, 12, 14
  of carbohydrates 10
Biochemical processes in aquatic
    systems 1
"Blooms" 14

Carbon monoxide 54, 56
  in human breath 60
Carboxyhaemoglobin 56
Cardiac glycosides 85
  estimation 88
  in milkweed 85
  in monarch butterfly 85
Chemical ecology 1, 84
Chemical defense of the monarch
    butterfly 84
  substances 84
Chlorophyll 4, 20
Chromatography 32
  gas 36, 38
  paper 72
  thin layer 32
Cineole 84

DDT 30
  in cigarette smoke 30
  in fatty tissues of man 77
  in human milk 77
  metabolism of 78
Digitoxin 86

Dissolved oxygen
  field measurement 95
  in productivity measurement 18
  in water 6
  meter 10
  solubility in water 11

Environmental chemistry 1
Energy flow in aquatic system 7
Erucic acid 74
  estimation in bread 75
Eutrophication 14
  indices 16

Field survey 91
  dissolved oxygen 95
  sample collection of storage 93
Food additives 69
  benzoic acid 69
  colours 69
  sorbic acid 69
  sulphur dioxide 69
Food additives and contaminants 62
Food colours in jellybeans 72

Gas chromatography 38
  column efficiency 39
  electron capture detector 77
  of DDT in human milk 80
  of fatty acid methyl esters 75
  of hydrocarbons 41
  retention time-carbon number
    plots 40

Haemoglobin 56

Insecticides
  carbamate 31
  DDT 30
  identification by TLC 33
  in cigarette smoke 30
  inorganic 32
  organochlorine 30
  organophosphorus 31

Juglone 84

LD_{50} and LC_{50} 43
Lead
  in gasolene 51
  in household paint 51
Light and dark bottle experiment 18

SUBJECT INDEX

Monarch butterfly 85
  lavae 87

Nitrate in water 14
Nitrite in water 14
Nitrogen oxides in air 54
  estimation 58
Nutrient enrichment 14

Objectives of the course 1
Organochlorine residues
  gas chromatogram of 81
  in human milk 79
Ouabain 86
Oxidants 57
  estimation 58
Ozone 54

Peroxyacetylnitrate (PAN) 55
Petroleum hydrocarbons 36
  gas chromatography 41
Photochemical smog 55
Phosphate in water 14, 17
Phosphorus in water 17
Photosynthesis 4, 14, 19
Primary production 6
Productivy in water 18

Reports on activities 2
Respiration 4, 19

Salinity - specific conductance
    conversion 98
Septic tank 26
Sewage
  biological filter 27
  composition 22
  degradation 23
  farm 24
  lagoons 24
  primary treatment 24
  secondary treatment 26
  tertiary treatment 27
  treatment 22
Solar radiation 6
Sorbic acid in soft drink 71
Stream flow 98
Stream pollution 2, 92
  presentation of report 99
Sulphur dioxide
  in air 54
  in soft drink 70

Theoretical oxygen demand 8
Thin layer chromatography 32
  of alflatoxins 67
  of insecticides 33
Threshold lethal concentration 49
Tolerance limit 43
Toxicity
  classification of 44
  LC_{50} and LD_{50} 43
  of copper ions to aquatic
    organisms 43
Toxic substances in the
    environment 1

Water quality experiments 10, 95
  ammonia 97
  available chlorine 97
  chlorophyll 20
  conductance 95
  oil and grease in sediments 95
  pH 95
  phosphorus 17
  stream flow 98
  suspended solids 95
  temperature 95
Water sampling 9
  and storage 93